THE ROLE OF PERCEPTION
IN SCIENCE

Charles Noël Martin

THE ROLE OF
PERCEPTION IN
SCIENCE

Translated by
A. J. POMERANS

HUTCHINSON OF LONDON

HUTCHINSON & CO. (*Publishers*) LTD
178–202 Great Portland Street, London, W.1

London Melbourne Sydney
Auckland Bombay Toronto
Johannesburg New York

First published in France under the title
Les vingt sens de l'homme devant l'inconnu

First published in Great Britain 1963

*This book has been set in Baskerville type face. It has
been printed in Great Britain by The Anchor Press,
Ltd., in Tiptree, Essex, on Antique Wove paper.*

Contents

Man, peeping at the Universe through only a few tiny windows—his senses—catches mere glimpses of the world around him. He would do well to brace himself against unexpected surprises from the vast unknown; from that immeasurable sector of reality that has remained a closed book.—MAURICE RENARD

Preface

Every scientist and humanist must pause and take stock of his knowledge at least once during his lifetime—and this book is the result of such a pause.

Modern science is advancing with terrifying speed, and it is only by consolidating newly acquired scientific knowledge that we can hope to plan the future wisely.

The reader must not expect a final answer to any of the problems discussed. The book is meant as a map and not as a mentor. Each chapter can do no more than touch on some aspects of its subject, and even these aspects are shown to be little more than fleeting reflections of our present knowledge.

Thus, I would like the reader to consider my book as outlining a general, provisional, approach rather than as an attempt to present final truths. For when we compare our knowledge with our vast ignorance it proves to be no more than a small child's understanding of the great adult world. Yet children do grow up if they are shown the right path. That path I have tried to pave in what follows.

Introduction

Though but a particle of dust exposed to the whim of space and time, man is a miracle in his own right. Constructed of inert matter, he yet lives, acts, and thinks.

Our universe, like us, consists of microscopic elements: atoms and molecules, obeying strange laws. Life, too, is built up of atoms, but it also involves mind and consciousness.

Our tools of perception, the senses, are our sole means of contact with environment. While we are a long way from understanding all the details involved in their functions, what we do know goes quite a long way.

Scientists have done much not only to explain our senses to us but even to 'extend' them with new instruments. These instruments have caused a veritable revolution in man's scientific outlook.

Moreover, our mind, also, is being scrutinized more closely. We are learning a great deal not only about its working but about its full potential and how to harness it to best advantage.

While our history is long, our future may be much

longer still. Either we shall be destroyed by our own foolishness or we shall learn to live life to its fullest, most prodigious extent.

These are the subjects we shall be discussing in what follows. While no man can ever hope to learn all there is to know, this book is an attempt to show what knowledge we have attained and how to cope with it. Only by fitting recent discoveries into a body of coherent and critical knowledge can we ever hope to sift the wheat from the chaff.

Analysis means dissecting our environment and ourselves as well. But analysis is not enough. What is needed now is a synthesis, whereby all the parts may be fitted into an intelligible whole. While modern science has every right to be proud of its discoveries, it seems to be splitting apart at the seams. Scientific progress may be compared to insect life. In its larval state the insect is pliable, plastic, capable of assuming a vast variety of forms, and of absorbing its needs from the environment. The chrysalis, on the other hand, is hard and constricted, incapable of growth, and short-lived.

If science is to survive it must become plastic again. While the actual future is anyone's guess, science (and mankind) must take a more positive view of life, and all that life entails, or perish. This book is an attempt to show why we have reason to trust in its survival.

I believe most sincerely that our actual knowledge is

but a dim candle, and that, as our awareness grows, we shall arrive at a view of the Universe far more brilliant than anything we now imagine.

While individual sections and chapters of this book can be read independently, there is an underlying thread which would be missed by so haphazard an approach.

Wherever I use the pronoun 'we', I mean mankind as a whole. Scientific knowledge is our collective heritage, hence few great names are mentioned.

This book is addressed to the general reader, and therefore eschews technical language. Those scientific terms which have had to be included are most carefully defined. I have deliberately guarded against being pigeon-holed as a specialist of any kind. This is simply the work of a human being passionately devoted to scientific research and to the study of the vast body of scientific knowledge which has been gathered by thousands of disinterested men over the centuries.

I

By Way of Beginning

IT IS always extremely difficult to express thoughts. Words and phrases are so many fetters by which our spirit is bound. Words are mere symbols of reality, and the written word is no more than a one-dimensional flow across the two-dimensional page of a three-dimensional book.

But symbols involve yet a fourth dimension: time. Symbols must be interpreted, and an entire book cannot be interpreted in a moment. It is time alone which makes possible the communication of thought, the miraculous contact between separate brains. This contact is often helped by emphasizing its purpose, and I have thought fit to begin my book by outlining its general plan.

The six chapters of this book are meant to sketch man's past scientific achievements and also many of his worst errors. In this preliminary chapter I shall discuss the essential role of language in all communication. While the word, and later the written word, have been the chief tools of man's primogeniture, language now threatens to become a brake on his future.

Chapter 2 deals with the Universe as we know it today, with its infinitely large galaxies, those vast congregations of stars from which we are separated by inconceivable distances. Light takes between one and two billion[1] light-years to travel to our present instrumental limits. Hence the Universe is never what we observe it to be at any given moment. Chapter 2 also deals with those infinitesimally small phenomena into which the Universe can be dissected. Atoms are the material of which molecules are built, and molecules go into the making of all matter. Their behaviour will be discussed in some detail.

Chapter 3 examines the same problems from a different angle, and shows how small is the part of the Universe that we really know. It also explains how atoms have been split into their constituent parts, and that atoms are not the fundamental particles they were once thought to be. What we consider to be fundamental particles today may well turn out to be complex particles in the future.

A discussion of 'energy' leads to the examination of life and living matter, and of 'time', that most mysterious of all phenomena. Relativity theory has made clear how intimately time is related to space, so much so that all space-travellers will have to take conscious cognizance of this 'new' dimension.

[1] Throughout this book, billion refers to 1,000 million.

Chapter 4 examines life more closely in the light of the preceding chapters. It shows how impossible it is to tell what life really is, and how our mind boggles at the task of having to explain itself. There is a section showing how life has developed from atoms, first into primitive cellular entities, and then into 'vast' organisms containing billions of cells; and another which describes how much all organisms are part and parcel of their environment.

Chapter 5 deals with our senses, and emphasizes how complex they really are. It is thanks to them only that we have discovered the world as this book tries to describe it.

Chapter 6 is an incursion into the future of science. We are about to cross the threshold of outer space, and it is imperative that we have some idea of what we may expect to find there.

As for the conclusion, it merely sums up the author's conviction that science, as we know it, is but the springboard towards an immensely promising future.

Language

We have seen that thought is confined by language, a fact of which we are normally unaware. We must cloak external events in words if we are to communicate with our fellow men at all. Most of our sense impressions can be translated in this way, though our feelings—a major part of our experience—frequently elude verbalization.

This is not the place to discuss the origins and physiology of language which has been studied by etymologists for more than a century, though with unconvincing results. Language, like so many other human faculties, may be analysed and classified, but it cannot be *explained*. Even less can we explain its origins, except by comparative physiological studies of animal behaviour.

While it used to be held that man alone was endowed with intelligence and the means of communication, it has recently been shown that every species has its own, probably very complex, method of signalling. Our earlier beliefs, like so many of our current ideas, have been shown up as purely anthropocentric—man-centred—prejudices.

Thus patient research has shown that bees communicate with one another by executing complicated dances in their hive; that such fish as eels and cramp-fish have special organs for emitting and receiving electric impulses, while other fish are known to communicate by sound; and finally recent work on crows has brought to light the amazing fact that, while crows have a 'language', European and American crows cannot understand one another.

Here, then, is a most promising field of research. The more we study our environment, the better we know ourselves.

But to return to the human language. Clearly our ability to join words into phrases, and to apply them first to sense impressions and then to abstractions, has had both happy and unfortunate consequences. Happy, because they have led to the development of societies and hence to civilization, to abstraction and hence to reason; unfortunate, because our vocabulary is inherently limited and thus tends to paralyse original thought.

Thought without words is extremely rare, since even meditation is usually no more than an inner monologue. Phrases, however incoherent, keep revolving in our minds, at least while we are fully conscious. But there is another way of thinking, that of the mathematician, who works with symbols which have no relation to language or empirical events. His is thought in the purest form, direct cerebration without the use of words. Creative artists, too, are 'mathematicians' in their own way.

Still, for most of us, language is a travesty, a distortion, of our inner truth. Often we have great intuitions, or flashes of genius, which words can only abase. Thus there are far more creative geniuses than we realize, for what we call masterpieces are masterpieces not only of creation but also of expression. It is not enough for great poets to be inspired, they must also have the means of conveying their inspiration to others.

Despite its shortcomings, however, language is our greatest tool. When it comes to conveying simple actions or ideas, this tool, handed down to us by our ancestors, serves us well. Simple language is like a record—once impressed on its grooves, the sound will never leave it.

But as science advances it increasingly comes across new phenomena which elude our vocabulary. New terms must be coined, often involving abstract notions or subtle distinctions. Since our language, and hence our views, are the result of millennia of human development, we have an innate resistance to all such innovations.

One way out of the difficulty would be to set up a new symbolic language based on mathematical logic. Alas, man feels safe within the circular wall of his dictionary, and prefers its security and very real rewards to the dangers of untrammelled exploration. Perhaps this is inevitable, perhaps we must outlive one stage of evolution before entering into another. Who knows whether coming generations will not look back at us with condescension, much as we look at ants who must rub antennae to tell each other whatever they do have to tell?

2

Man's Corner of the Universe

THE INFINITELY GREAT

BY MEANS of his perception and intellect, man has been able to construct a coherent model of the Universe in which he lives. Before discussing the details of this model, we must outline it in profile.

In the following chapters we shall therefore embark on a wild journey which will lead us from one infinity to another. Galaxies and protons alike take us to the limit of our power of comprehension, a limit which science is doing its utmost to extend.

The night sky has always fascinated mankind, so much so that for countless generations men have been gazing at the stars, first in sheer wonderment and then with ever-growing understanding.

Long study, patient calculations, and occasional inspirations have led astronomers and astrophysicists to the view of the cosmos that we shall sketch in the following

21

pages. The reader would do well to remember that this view is only provisional and that much work remains to be done.

Photons—the particles of light

Centuries of observation have given scientists some idea of the immense size of the Universe. In this idea, space can no longer be separated from time.

It is mainly through the study of photons that we have gained what knowledge of outer space we have. A photon is a particle devoid of mass, which is associated with all electro-magnetic radiation. Light-waves, radio-waves, X-rays, and gamma rays are all examples of photon radiation of different intensity. The photons associated with a radio-wave have less energy than those of a light-wave, and those associated with a gamma ray have considerably more. However, all photons have one common characteristic: they all travel through space with the same velocity, viz. at 299,792 kilometres per second.

The surfaces of all luminous bodies and of all radio-active substances emit electro-magnetic radiation of varying intensity. The sun is a case in point. It has a surface temperature of about 6,000°C., and therefore expels a tremendous amount of energy into space, where that energy travels with the speed of light. A very small part of this radiation is intercepted by cold bodies, e.g. the earth and the other planets, much as light from a candle is absorbed by objects in a room. The earth receives all sorts of photon radiation, ranging from Hertzian-waves to ultra-violet rays, which latter are absorbed some sixteen miles above sea-level by the ozone in the earth's atmosphere. This range of radiation includes the infra-red rays, which convey heat, and the entire spectrum of visible light from red to violet, with a maximum intensity between the yellow and the green, corresponding to the maximum intensity of solar radiation. The fact that our eyes can see the spectrum is a striking example of the way in which our senses are adapted to their environment. If our sun looked blue, like some other stars, we should most probably be able to see ultra-violet rays, rather than red. These facts will be discussed more fully in Chapter 5.

Unconscious messengers of time

When we consider the immense distances which separate us from even the nearest stars, and the fact that light from these stars takes time to reach us, we realize that the mere evidence of our eyes is most deceptive. Light, travelling at almost 300,000 kilometres per second, takes four years and four months to reach us from Proxima Centauri, the star nearest to our sun. Sirius, the most brilliant star in our sky, is 8·6 light-years, and the Pole Star fifty light-years from the earth. Moreover, these stars are comparatively near when we consider that the distance between us and that teeming multitude of stars of our Galaxy that can be seen only through telescopes must be measured in thousands and tens of thousands of light-years.

In other words, a photograph of any section of space is bound to give us a false picture, for no two stars are the same number of light-years from us. The light from one star may have taken 5,000 years to reach the lens of the camera, whereas that of its neighbour may have been travelling for 10,000 years. Neither star will have been in the exact spot shown in the photograph when it was taken. Moreover, during the interval between the light leaving the stars and reaching us other changes may have taken place. A catastrophe may have destroyed them and, for all we know, the stars which we see burning so brightly may have become extinct long ago. The Crab Nebula is a typical example.

A 'fossilized' Cosmic explosion

In the year 1054, the Chinese encyclopaedist Ma Tuan-lin recorded a sudden explosion of a star in the constellation of Taurus. The explosion of a star (i.e. the formation of a supernova) is the result of an abrupt change in its structure, with the consequent accumulation of so much energy that the star disintegrates within a few days. This phenomenon is so rare that we can expect to record it, in our Galaxy, only once every 500 years. When it happens as much energy is released in one month as radiates from the sun in twenty million years. Supernovae appear suddenly in the sky, become very bright, and then fade away very rapidly, to disappear completely from sight within only a few months.

The supernova of July 1054 has since been rediscovered: it is the Crab Nebula, a cloud of matter ejected from the original star, and hurled into space at 1,000 kilometres per second. The supernova was 4,000 light-years away from us, so that the actual date of the explosion, which was observed in 1054, must have been about 3,000 B.C., and the Nebula (which we observe today) must have been formed in about 2090 B.C. Moreover, the present state of the Nebula will be observed only when another 4,000 years have passed.

We must realize, then, that our impressions of the starlit sky are completely misleading. We can never have any idea of the real state of the Universe, since what we actually observe has long since ceased to exist.

On the edge of a spiral

A further example of the same difficulty is provided by the Galaxy of which the earth is an insignificant unit —the Milky Way.

The Milky Way is made up of hundreds of billions of stars, each similar to our own star, the sun. Some are brighter than our sun; some much less bright; some fluctuate, and others emit light sporadically. They also differ greatly in size, and many form binary or multiple

25

systems, the members of which revolve in mutual orbits.

Our view of the Milky Way is due to our peculiar perspective. In reality the Milky Way is a spiral disc, with a diameter exceeding 100,000 light-years. This disc is very flat (5,000 light-years) and has a thickened nucleus in Sagittarius, which is hidden from us by a mass of luminous stars. From this nucleus great spiral arms branch off. Our sun is situated between two of these arms, so far from the centre as to appear almost at the periphery of the system.

Today scientists are trying to determine the exact shape of the Milky Way, mainly by investigating the wave-lengths emitted by inter-stellar hydrogen. Now, the waves in question have been travelling for distances up to 80,000 light-years before they reach us, and as our whole system completes one revolution every 200 million years, our direct observations must be corrected. Clearly, the spiral galaxies have not remained rigid during that time, and there have been profound changes within the Milky Way, which have affected the distribution of the stars within it.

A crowd of individuals

At this point it might be as well to define certain terms, which may otherwise confuse the reader. The word 'Galaxy' refers to the Milky Way, while 'galaxy' refers to any system of stars in the Universe. Another term for galaxy is 'island-universe'; 'Universe', on the other hand, is an all-inclusive term, embracing our Galaxy and every galaxy past, present, and future.

Modern astronomers know an immense number of island-universes, for some hundreds of millions of these, each consisting of between 100 and 200 billion stars, are within range of our telescopes. Only one of these countless galaxies is near enough for us to see with the naked eye. This is the Nebula of Andromeda, which in size, and in the number of stars which it contains, resembles the Milky Way, from which it is separated by two million light-years.

Although technical strides in recent years have been tremendous, our means of astronomical observation and analysis are still quite inadequate to solve all the problems facing us. Moreover, as our knowledge increases, so do these problems multiply.

The galaxies contain two kinds of stars, classified as population I and population II respectively. Although these two populations overlap, we know that they differ in age and history. Galaxies differ widely in shape: the special form of the Milky Way is unique. Galaxies have satellites, e.g. the Megallanic Clouds, two oval cloud-like patches of nebulous light hovering at the edge of our Galaxy, and visible in the Southern Hemisphere.

All these complications confuse our ideas about island-universes. Still, our greatest problems arise from the vast distances involved, for many galaxies are so far away that they completely elude our observations. Once again, our vision has failed us.

Looking across aeons

The need for relating time to space is felt by all astronomers. The time taken by light to reach us from the most distant galaxies must be measured in many millions of years. Thus giant telescopes and radio-telescopes record light- and radio-waves which began their journey a billion years ago. In the near future radio-telescopes will probably replace optical instruments altogether, simply because radio-waves travel farther than light-waves. Radio-telescopes will be able to track down radiation emitted four to five billion light-years ago.

But even if our analysis of electromagnetic waves from outer space is correct, the resulting knowledge is bound to be distorted, since our instruments merely tell us what the galaxies were like five billion years ago. In other words, while we can receive signals from distant stars, none of these signals can tell us anything about the *present* state of the Universe. A correct view of the Universe must combine knowledge of vast distances in space with a study of vast intervals in time.

Any anthropocentric approach to the Cosmos, i.e. all ideas that things happen in a certain place, at a given time, are bound to be false. We must cease to assume that what laws we have established to hold on earth are bound to hold in outer space as well. All the evidence goes to show that whatever the Universe may be it is certainly not cast to a mould, i.e. it is not a system of homogeneous units organized according to similar patterns.

Is the Universe a soap bubble?

As our telescopes reveal more and more complex phenomena, we begin to realize how limited our ideas really are. All attempts to discover the 'frontiers' of the Universe are doomed to failure from the start. Far from observing any point where stars cease to exist, we find that the greater the range of our instruments, the greater is the multitude of stars they reveal.

At this point we must consider the problem of the expansion of the Universe, which is closely connected with the problem of infinity. During the last thirty years much has been written on this question, though the question itself has been wrongly posed. Consciously or unconsciously, man's mind is always anthropocentric. Since it is impossible for us to form any coherent idea of infinity, we try to produce a finite plan of the Universe instead. Thus it has been suggested that a reasonably exact model of the Universe would look like an infinitely large gas-filled balloon, each molecule of the gas representing a galaxy. Now, relativity considerations show that such a balloon must contract or expand with time: hence

there arose the idea of an expanding universe, corrobora-
ted by the fact that light from distant sources shows a shift
to the red, i.e. it shows that the observed sources are
moving away from us. Be that as it may, our anthropocentric mind continues to boggle at the idea of a balloon,
which, being infinite in size, can yet continue to expand.
It cannot grasp the fact that the farther away a galaxy
is, the faster it travels away from us. Yet there is no doubt
that while the region of most distant galaxies seems to
expand at a rate approaching that of light, galaxies four
or five hundred million light-years from us appear to be
retreating at a speed of 150,000 kilometres per second,
i.e. only half the velocity of light.

Our sight is short

But even if this phenomenon is an observable fact, we
have no right to infer that the rest of the Universe is
receding from us, for our observations are purely relative.
Recent discoveries have shown that there may be other
explanations, and that the balloon theory does not necessarily fit the facts. Thus intersidereal and intergalactic
space is now known to be much less empty than it was
formerly thought to be. It contains dust and cosmic gases
whose physical effects might perfectly well account for
the observed shift to the red on which the balloon theory
is built. This does not mean that there is no expansion of
the Universe, but it does mean that that expansion is

much less general and much less basic than scientists believed. What we see may be only a minute part of the Universe, subject to fluctuations in space and time, a mere aspect of a larger system that is far beyond our comprehension.

Clearly all scientific progress is based on extrapolation: on constructing new hypotheses from old and established theories. But we cannot extrapolate spatial theories into time-space hypotheses.

Consequently, we must look at all our knowledge with a great measure of humility. The sum total of our experience is insignificant, and all our methods are inadequate. Thus we can never hope to grasp the immensity of the Cosmos, though patient observation and diligent work may help us to solve some of its mysteries.

STARS AND PLANETS

So far we have spoken mainly of luminous stars. Now, there is no reason why we should assume that things do not exist simply because we cannot see them, or why we should think that all the stars in the Universe must glitter.

Stars consist of an immense number of atoms making up a hundred different elements, all mixed in different proportions. It is a striking fact that the spectral analysis of light from the stars reveals the presence of all (and none but) the chemical elements found on earth.

Nuclear furnaces

The release of energy radiated by the stars is the result of reactions among their atomic nuclei. The temperature near the centre of a star is extremely high, at least ten million degrees Centigrade, and in the case of certain very 'hot' stars as much as two billion degrees. Atomic nuclei are so active, and move so fast at such high temperatures, that collisions between them become very frequent indeed. Nuclear fission then takes place, exactly as it does in a cyclotron, with a consequent change in the number of nucleons of the nuclei, and the transformation of an element into another, higher element. In the case of the light elements (hydrogen, helium, lithium, etc.) this transformation releases surplus energy which travels through matter and into space. Einstein's calculations have shown that the sun releases four megatons of energy per second.

All stellar furnaces are in a dynamic state. On a star which consists of from 80 to 90 per cent hydrogen, this gas is gradually transformed into helium and other light elements. During that time the temperature becomes higher still, and more and more complex and heavy elements are formed by nuclear reaction. The other elements of the periodic system begin to appear and to evolve towards such elements as iron, copper, tin, and the rare earths. As the temperature continues to rise, further reactions take place, involving alpha particles (helium nuclei), which may be called the ashes of the first stages of stellar evolution. This, in brief, is how the heavy elements are built up.

Exploding stars remodel the Universe

At a given point the tempo of stellar evolution becomes so accelerated that catastrophes occur. When that happens we have novae, stars which emit a sudden concentration of energy, and supernovae, which explode in one great burst of light, scattering interstellar dust through cosmic space. The total mass of this dust is of the order of that of all the stars in a given galaxy.

Now it is becoming increasingly certain that stars were originally accumulations of cosmic dust. In fact photographs have shown that young stars, still in the process of being formed, arise in regions particularly rich in dust.

Perhaps, then, the evolution of the Universe is a process of continual rebirth. Stars take shape, evolve, create complex elements, explode, and by exploding broadcast their dust into space. The dust then collects to form a new star and so on.

Our understanding of this process is, however, far too sketchy, since astrophysics, though a promising branch of science, is still in its infancy. All we can say is that we *think* that stars have existed for five or six billion years. The expression 'have existed for' may be misleading. We observe the state of the Universe at our own particular

C

time and then try to fit it into our own uniform and recti-linear time-scale, for better or for worse. In any case, research of radio-active disintegrations of such elements as uranium indicates that the crust of the earth is at least four billion years old, and that the stars must be at least as old as the earth. More we cannot say with certainty.

Billions of planets

Besides the nine main planets of the sun, our solar system contains tens of thousands of other bodies, ranging from gigantic rocks over 100 miles in diameter down to mere pebbles. Other stars, too, have their planets, and when we consider the number of known stars the total number of planets becomes incalculable. We may take it that these planets were originally part of the star round which they have kept revolving ever since they split off from it.

To say that the sun's satellites are the results of ex-tremely rare 'accidents' is but another instance of anthro-pocentric thinking. Even if the necessary conditions are unusual, the Universe contains so fabulous a number of stars that such 'accidents' are bound to arise.

For some years now scientists have inferred the pres-ence of dark bodies in the neighbourhood of the nearest stars. These planets are some ten times as large as Jupiter, the greatest planet of the solar system. Now if the nearest stars have large satellites it is more than likely that they have smaller ones as well.

The existence of such planets suggests the possibility of life even outside the solar system. However, before we go into this question more thoroughly, we must first examine the physical conditions under which life can flourish.

All sorts of worlds

We have much to learn from the planets of the solar system. Each planet, from Mercury to Pluto, has a characteristic temperature and absorbs specific radiation. The sun emits a limited spectrum, which has a maximum intensity in the region of yellow. That is why the sun looks so golden to us. From Mercury, the sun looks like a great blazing star emitting light of unbearable intensity, and hot enough to melt lead. From Pluto, the sun looks a fairly dim star that cannot raise the temperature by more than a few degrees above absolute zero (i.e. 240°C.). This temperature is so low that gases on Pluto must be completely solidified, unless heat is generated by internal radio-activity, as it is on earth. Now life is intimately bound up with temperature, and can exist only where temperature conditions permit.

Life is also affected by the force of gravity. All bodies are attracted to other bodies according to the quantity of matter which they contain. The first men to land on the moon will find that they have only one-sixth of their terrestrial weight. If they could reach Jupiter their weight would be two-and-two-third times greater than it is on earth. Those large planets in outer space which we have mentioned must have a much greater pull still, while gravity would be negligible on, for instance, the minor planets (asteroids) which circle the sun between Mars and Jupiter.

Life is also affected by other surface conditions. The moon, for instance, is constantly bombarded by meteorites which cause great surface disturbances, as do the temperature differences of up to 200°C. between the lunar day and night (fourteen days). The resulting expansions and contractions of its surface have, in fact, caused this surface to become pulverized to dust to a depth of several yards. The surfaces of Jupiter and Saturn are probably in a state of constant flux, for gaseous and liquid layers are known to keep intermingling.

Some planets, such as the earth, have a great deal of water; others, like Mars, have very little. The moon is an example of a planet that is completely arid. Venus and the earth have an atmosphere, the moon has none, and that of Mars is highly rarefied. The composition of the atmosphere of a given planet may differ greatly from that of any other. Thus the atmosphere of Jupiter consists mainly of hydrogen, helium, ammonia, and methane.

Our imagination boggles

If so much diversity is found within our own solar system, how much greater the differences between two systems!

There are blue stars which emit waves of ionized X-rays, and young planets which emit α, β, and γ rays. Some planets rotate so fast that the force of gravity is much greater at their poles than at their equator, and others, again, have atmospheres of the most extraordinary composition. No matter how much free rein we give to our imagination, even our wildest fantasies must be fulfilled somewhere in the Universe.

Moreover, we must be careful not to be too anthropocentric in our definition of the conditions under which life can exist. Thus, though the pressure six miles below sea-level is so great that water will penetrate the thick walls of a glass vessel, some animals flourish under these conditions. On earth life is confined to the surface layer, but on other planets life may perfectly well exist in the depths of solid matter or even in quite unsuspected states.

Clearly we shall learn of such conditions only when space-travel has become a practical reality.

THE INFINITESIMALLY SMALL

As we turn from the study of vast spaces to that of minute particles, we are transplanted into a world far more infinite still than that of astronomical dimensions. Beyond a certain point we again discover a prodigious emptiness, comparable to that of intergalactic and interplanetary spaces. It seems, then, that matter and life are contained between two voids—one at either end of the scale.

A detailed discussion of elementary particles, which strike us as the ultimate bricks of reality, will be given in Chapter 3 of this book. Here we are merely concerned to show how these particles combine into complex groups, creating first atoms, then molecules, and finally crystals and organic substances.

The emergence of atoms

Everyone knows that living organisms consist of arrangements of individual cells. This is plain to all who have ever examined a living tissue under a microscope. All the complex phenomena we see are the result of cell combinations, of which there are an immense number. However, this complexity must not blind us to the fact that living beings, no matter how intricate, are all derived from the division of one primary cell. By drawing on inert matter from its environment, this tiny primary unit has grown larger and become organized and differentiated.

Organic cells suggest that inorganic substances, too, can be divided and subdivided down to a basic unit, common to all inert matter.

In the last 200 years scientists have looked into this problem with great success, so much so that they can now analyse most structures—inert and living. The atom, the unit of which all matter is built, has finally been tracked down.

Two thousand five hundred years ago, when Greek philosophers formulated the atomic theory of matter, they arrived at it by a process of reasoning, rather than by scientific research. This fact is so well known today that we need do no more then mention it here. Nowadays it has become clear that all matter—liquid, solid, or gaseous —is made up of combinations (molecules) of atoms. Thus, when sugar dissolves in water, what happens is that the sugar molecules intermingle with the water molecules. These molecules never lose their identity, for, though the water assumes a sweet taste, the original molecules can be separated out.

The history of corpuscles, grains, and particles

For the past 200 years scientists have striven to isolate the basic unit of matter. This work has become greatly accelerated since the turn of the century.

To begin with, chemistry prepared the way, by showing that elements combine in simple arithmetic proportions. By 1880, when most of the elements had been discovered one by one, it had become clear that all the many forms which nature assumes are composed of fewer than a hundred elements. Moreover, these elements could give rise to a host of forms not hitherto discovered on earth.

Later on, physics set the atomic theory of matter on a firmer foundation. Electricity was shown to have a granular structure and the electron, the ultimate grain of electricity,

39

was isolated in 1895. With that discovery all previous atomic theories had to undergo radical revision.

Next came the discovery of radio-activity. It was found that certain of the heavier elements, such as polonium, radium, and uranium, are capable of emitting radiation other than that found in the solar spectrum. These elements are unstable, i.e. subject to transmutation effects. For instance, radium generates the gas radon, which turns into polonium, and finally into lead. Thus certain elements can change into others, for radium, radon, polonium, and lead are all distinct elements of atomic numbers 88, 86, 84, and 82 respectively. Moreover, radium itself is derived from uranium, whose atomic number is 92.

Three important problems are posed by these discoveries: How does this radiation come about? Why does spontaneous transmutation occur? Do all radio-active elements follow a similar disintegration pattern?

From 1900 onwards the study of radio-activity forced scientists to revise all their past concepts once again. By 1920 a complete atomic theory had been built up, and from then on the study of physics progressed by leaps and bounds.

Painstaking analysis has, bit by bit, laid bare the vast potentialities hidden within each tiny atom. One millimetre would cover tens of millions of atoms placed end to end—hence the impossibility of observing dissolved sugar molecules in water even under an electronic microscope, which has a magnification of up to 200,000 diameters. To see atoms, we should need a magnification of from two to five million diameters. But even such powers of magnification have recently been attained, when photographs were in fact taken of individual atoms, or rather of their contours. For atoms are practically devoid of contents, since they consist of a swarm of electrons revolving at relatively vast distances about a small central nucleus. The construction of an atom is very similar to that of the solar system, in which a number of planets revolve about a central star. We may form some conception of the 'emptiness' of an atom if we appreciate that if a barrage balloon, representing the nucleus, were placed in the middle of London, the orbit followed by the electrons would pass through Dublin, Glasgow, and Edinburgh.

Electrons are important, not only because they are the source of all electric and electronic phenomena but also because they simplify the chemists' work. Modern chemistry deals largely with atomic electron distribution and energy exchanges between peripheral electrons. The greatest number of electrons found in any atom is 101, and fall elements are numbered according to the structure of their atoms.

Carbon, for example, has the atomic number 6. There are 12,000 billion billion carbon atoms in a diamond weighing one carat, and the nucleus of each of these atoms is surrounded by six electrons. At the upper end of the scale the uranium atom has a nucleus surrounded by ninety-two electrons.

The atomic scientist—the alchemist of the twentieth century

Nowadays certain elements can be changed into others with great ease. To do so we must act upon the nucleus, the real seat of the atom. Now the structure of the nucleus is disconcertingly simple—all nuclei consist of neutrons and protons alone, combined in simple arithmetical proportions. Despite its apparent simplicity, the nucleus, which has become the chosen field of study of nuclear rather than atomic physicists, can assume a host of guises.

What we have said about the nucleus is, of course, a crude oversimplification. Whole volumes would have to be written to do justice to all the recent discoveries of nuclear scientists. What is quite clear, however, is that, because of the advances made in the last generation, scientists have succeeded in re-creating many of the elements which have disappeared from the earth because of relatively quick radio-active disintegrasion. Thus elements 93 to 102 can now be synthetised, including an isotope of plutonium (94), which has a half-life of 25,000 years. Only a sixteenth part of the plutonium now stored in atomic piles and bombs will still exist a hundred thousand years from now.

Molecules—the minute bricks of our mighty Universe

We shall conclude with a brief explanation of the way in which atoms combine into molecules. Before doing so,

42

however, we must consider an idea which has not, so far, been mentioned—the idea of geometrization. By geometrization we mean the way in which we commonly represent molecular chains of atoms.

When atoms combine to form molecules, their relative positions are as important as their number, for their spatial arrangement determines molecular properties.

This phenomenon gives us an important clue to the mystery of nature and life. While we have known a vast number of atomic combinations for some time, we have for too long ignored their inner arrangements. Yet many of the problems posed by certain molecules (those of cellulose or rubber, for example) are solved by the simple examination of the atomic distribution within them.

Groups of molecules form crystals, and here again geometrization plays a part. Life, too, depends on the way in which organic molecules are arranged, for certain viruses are macro-crystals, i.e. large molecules, recurring in great numbers in a vast geometric pattern. All particles of matter smaller than one ten-thousandth of a millimetre have a crystalline structure.

So much for our quick sketch of the inner properties of matter. Between simple atoms and complex organisms there is a range of increasingly complex structures, combined in various ways. The way in which they combine determines all chemical and physical properties.

At present the full significance of the existence of geometrical patterns in nature continues to elude us.

3

The Envelope: Space-Time

THE phenomena we have been discussing all fit into a framework of three-plus-one dimensions. Within that framework of space-time, matter and energy are apparently two interchangeable aspects of one and the same phenomenon.

The following section is divided into four parts: Space, Time, Matter, and Energy. Time is dwelt upon at length because it is the most fascinating of all the mysteries surrounding us. The pages devoted to Matter contain a discussion of the most recent nuclear discoveries, the cornerstones of all future developments in physics.

Man has an intuitive idea of space, and his idea strikes him as most reasonable. With it, he has managed to travel not only across the earth's surface but even through the air. No wonder, then, that he imagines that outer space is bound to fit in with his preconceived notions.

Travelling in three dimensions

We generally limit all our movements to a single dimension. Roads and railways are in the form of straight lines. When we travel by rail it is linear geometry which governs our movements, and the same is true, more or less, of long-distance travel by road. Even an aeroplane, which seems to have conquered the third dimension, generally flies from one aerodrome to the next in a line parallel to the earth's surface. Only the stunt pilot and birds have a real mastery of all three dimensions, for they dive, climb, beat to windward, and turn in any direction. Normal movements, however, are restricted by the pull of gravity.

Now, there will be no gravity in a rocket moving in an orbit through space. Travellers on space-ships leaving the earth far behind will be subjected to external forces only while the space-ship is accelerating or decelerating. Particular directions will no longer exist. There will be no up, down, and no compass bearings. Bodies will be free to move at will in any direction, but this freedom does not seem to lead to any serious biological difficulty. This may be inferred from recent Soviet experiments with dogs, the American experiment with the chimpanzee Ham, the earlier tests with mice and other small animals and, far more recently, the experiences of Russian and American astronauts. In far outer space, however, man may experience extreme mental confusion and nausea. Model

45

experiments have shown that many subjects become unable to add up two simple numbers, each of only two figures.

Man's organism is adapted to a particular sort of space: a continuum with three dimensions intensified locally by a strong gravitational field. This point can only be understood in connection with Einstein's Generalized Theory of Relativity propounded in about 1914, or rather in connection with his Unified Field Theory. During the latter part of his life Einstein was working on a single theory which would account for all known physical forces. We shall see what these forces are on page 65. At present it is enough to state that, according to that theory, gravitational, nuclear, and electromagnetic forces are closely related, are mutually derivative, and form part of one entity.

When the earth makes its own bed

This vision that all forces can be reduced to obey a single law has been the greatest driving force behind Einstein's later work. Einstein believed that he had discovered the connection between gravitation and electromagnetism. He was not sure, however, whether his four field equations also governed the behaviour of the nuclear field, and his doubts were particularly marked just before he died. Einstein's unified theory is, in fact, no more than one of many possible hypotheses arising out of concerted efforts to synthesize relativity and quantum theory. Heisenberg's contribution in 1958 was among the most important steps in this direction.

Einstein's achievement in combining gravitation with geometric space was nevertheless momentous. According to this theory, mass, that is a given quantity of matter, will invariably produce a curvature of space. As a result, any heavenly body is located in a depression of space-time. Now if one heavenly body passes near another its trajectory is affected by the shape of space all along its path. Its direction will suddenly be altered under the action of an external force, i.e. the curvature of space induced by the second body.

We may illustrate this point by considering the case of an artificial satellite. The earth produces a depression of space-time, much as a small ball causes a thin sheet of rubber to sag. Now the artificial satellite may be likened to a tiny ball, launched so as to run along a 'rubber sheet' round the earth. If its original speed is great enough the ball will keep circling along this depression, gaining momentum every time it falls. Obviously friction will finally cause the ball to lose speed and drop, and this is precisely what could happen in the outer layers of our atmosphere, i.e. at a height of up to 250 miles. However, under ideal conditions, i.e. where friction is completely absent, the ball will go on revolving indefinitely. These ideal conditions are enjoyed by the moon, which has been following a path governed by the earth's space-time

47

curvature for billions of years. In the same way the earth describes an orbit determined by the sun. Gravitational attraction and the deformation of space are one and the same thing.

Mysterious space-time

Einstein's success in explaining so complicated a phenomenon as gravitation has prompted many other scientists to continue his work and to develop his unified field theory.

One of Einstein's greatest achievements was to show that space cannot be separated from time. The two are, in fact, so closely linked that they must be studied together. Mathematical considerations have shown that, while real variables can describe the three dimensions of space, time can be described only by the introduction of complex variables containing the factor i, i.e. the square root of -1. For that reason, space-time is said to have three plus one, rather than four, dimensions, thus emphasizing the peculiar character of time.

We should like to emphasize that the irreversibility of time is a mere hypothesis, based on the observation of macroscopic phenomena. Now, nuclear processes may very well run in reverse and their behaviour may be governed by laws which reflect a complete space-time symmetry, i.e. they can run backward in time. Hence intra-atomic space would have a very different structure from ordinary space, for while intra-atomic space would have three (or more) plus one dimensions, ordinary space (in which time is irreversible) has but three plus a half dimensions.

The introduction of a fractional dimension is due to the fact that, ordinarily, time moves in only one direction. Known macroscopic phenomena can move only into the future and never towards the past.

Certain modern theories go further still, and speak of five, six, or seven dimensions of generalized space. However, to my mind, generalized space is a mathematical rather than a physical reality. I believe that theorists are wrong to assume that *the whole* of nature can be described by *a single aspect* of current mathematical thought. Anthropocentrism is at fault, once again. There is a much better approach, which occurred to me at the very start of my work on theoretical physics. In it we assume that abstract concepts reflect physical facts. This view will be developed in a subsequent work.

Is death a passport to another dimension?

It must be remembered that however fascinating our notions of space and time may be, they are still extremely rudimentary. Man is only just beginning to delve into these mysteries, and much work still remains to be done. We know that nuclear space appears to have properties that are characteristically different from those of ordinary space. In fact, it seems likely that these properties are simply due to the space-time structure of the nuclear inter-spaces. The mathematical study of this phenomenon is an urgent task for the future.

Who knows what this work may reveal? It will be particularly interesting to see how the results will affect boundary conditions. For instance, we should like to know what would happen when the boundary separating a universe of $3+1$ dimensions from another of $4+0$ dimensions is crossed. Clearly the speed of light must approach zero precisely at that boundary, i.e. where we pass from a world with a time dimension to one in which time plays no part. Now is not death tantamount to a freezing of time? For when we die our 'world-line' is suddenly cut, and we may be said to have crossed a geometrical boundary, beyond which time does not exist.

It is possible that the vague idea of eternity, to which mankind has held fast so tenaciously throughout history, is the result of an obscure intuition according to which we pass out of this life to continue in an unknown dimension. Ordinary space-time would then be no more than one aspect among a host of possible ones. Our life on earth takes place between two known boundaries: our birth and our death. What if these boundaries were not as final as we think?

Imagination, aided by mathematical intuition, suggests a host of revolutionary possibilities, which, so far, no one has bothered to investigate. Just as geometrization plays an essential part in the microcosm, so must it also play a part in our life and thought, in personal and in cosmic experience.

Obviously these speculations concern matters quite beyond our present understanding. I have merely tried to indicate what vast revelations are promised by future research into the physico-mathematical aspects of space and time, and what tremendous new problems we may have to face.

THE STUFF OF ATOMS

As we saw previously all matter can be reduced to atoms. The stars, cosmic dust, the planets, living creatures, water, and air, are all constructed of vast assemblies of atoms. Atoms are the bricks from which all forms of matter are constructed. Now matter is, in fact, so closely related to energy that the two are often confused. Energy will be dealt with in a later chapter; for the moment we shall concentrate on matter, in its atomic form, alone.

What isotopes are

There are over a hundred different types of atom. Atomic numbers, and hence the elements which they represent, range from hydrogen (atomic number 1), to nobelium (atomic number 102), which was produced by atomic

physicists in 1957. We have seen that the difference between one type of atom and another is purely arithmetical.
So far, three constituent atomic particles have been
isolated—the electron, the proton, and the neutron.
Protons and neutrons are found in the central nucleus
round which the electrons revolve. The nucleus is one ten-
thousandth the size of the orbit of its peripheral electrons.
There are the same number of protons and electrons in any
one atom, and this number is, in fact, the atomic number
of the element concerned. For example, oxygen has eight
electrons and eight protons. The nucleus contains a
variable number of neutrons, but this variation does not
affect the identity of an element, though it is taken into
account when, for instance, we speak of atomic weight
(the multiple of the masses of protons and neutrons taken
to be approximately equal) rather than atomic number.
Thus we have O-16 (i.e. oxygen containing eight neutrons
and eight protons) and O-17 (i.e. oxygen with nine neutrons
and eight protons. O-18 stands for oxygen with ten
neutrons and eight protons. These three types of oxygen
nucleus (called 'isotopes', because each represents oxygen)
can all be found in nature. The oxygen we breathe is a
mixture of all three in definite proportions (99.759 per
cent of oxygen 16; 0.037 per cent of oxygen 17; and 0.204
per cent of oxygen 18). Isotopes have approximately the
same chemical properties, though their physical properties (viscosity, specific heat, conductivity, latent heat, etc.)
may differ.

Radio-active alchemy

There are isotopes of every natural element, including radio-active isotopes, e.g. O-15 (eight protons and seven neutrons), and O-19 (eight protons and eleven neutrons). These two isotopes have unstable nuclei, i.e. nuclei which emit electrical energy spontaneously, thus altering the ratio of neutrons and protons. Thus half the oxygen 15 (which is produced in cyclotrons or other particle accelerators) turns into nitrogen within two minutes, while oxygen 19 changes into fluorine (atomic number 9) after a half-life of a mere thirty seconds.

Clearly, therefore, matter, for all its basic simplicity, assumes a host of complex forms. While electrons characterize the chemical nature of the different elements, the existence of isotopes and their stability are all due to the action of protons and neutrons alone.

That being so, what part of the atom may be said to contain matter? We believe that matter is mainly concentrated in the nucleus, although a very small proportion of matter is found in the electrons. Thus, while the oxygen nucleus weighs $2 \cdot 66.10^{-23}$ gm., i.e. $26 \cdot 6$ millionths of a billionth of a billionth of a gram, the eight electrons together weigh less than five thousandths of even this minute amount.

Even the heaviest atom, that of natural uranium (element 92), weighs no more than four ten-thousandths of a billionth of a billionth of a gram.

Protons and neutrons as an inseparable couple

Now that we know how matter is distributed inside the atom, we can look at atomic particles more closely.

Though atomic physicists have been studying these particles for thirty years, they are still very far from having solved all the problems involved. The history of nuclear physics is a perfect example of the advances, setbacks, unexpected pitfalls, and revolutionary changes which have always beset scientific research.

The physicists' first job was to determine how many types of particle went into the making of an atom. Experiments from 1933 onwards have shown that there are at least three distinct types, i.e. the electron, the proton, and the neutron, all of which are stable in the *normal* state. The electron, which plays a part in all electrical and electronic phenomena, weighs only a billionth of a billionth of a billionth of a gram. It carries an electrical unit charge of −1. Its diameter is a thousandth of a billionth of a centimetre.

Models for poor mathematicians

The proton and the neutron, which make up the nucleus of atoms, are collectively called 'nucleons'. They appear to have very similar properties, and weigh respectively 1,837 and 1,839 times as much as the electron. The proton carries an equal, but opposite, electrical charge to the electron. The neutron has no electrical charge; indeed, it is called 'neutron' for that very reason. The entire nucleon occupies no more than between fifteen and twenty thousandths of a billionth of a centimetre.

These figures are too small to be grasped by ordinary thought, and we are forced to use comparisons if we are to think of them at all. For example, it is quite within our understanding to say that a light-wave occupies a little less than a ten-thousandth of a millimetre. Now an electron is one millionth of the size of this wave. The nucleon is 2,000 times smaller than the space occupied by an electron. So far so good. Unfortunately, these particles which we have called 'elementary' are not really so, for they contain even smaller sub-particles.

Intra-nuclear invasions from outer space

From 1937 onwards all atomic theories were shaken by the discovery that nuclear collisions in outer space produce quite unsuspected particles. Atoms in outer space are shorn of their peripheral electrons as a result of the rapidity with which they travel, and the resultant protons and neutrons hit the earth's upper atmosphere much as if it were the target in a cyclotron. The high-energy particles collide with the oxygen and nitrogen nuclei of the air, and the impact is comparable with a catastrophe, though on a nuclear scale. Further nuclei are split up and the resulting free protons and neutrons scattered. It was among these fragments that the new particles were first detected.

That particles could leave the nucleus at all indicated that the nucleus itself was not as simple as scientists had thought it to be. Thus beta-radiation, i.e. the emission of a negative, or short-lived positive [*sic*] electron, posed the complicated problem of how a nuclear collision could liberate particles which apparently occupied a larger volume than the nucleon itself. This problem is still being investigated.

The particles produced by cosmic radiation are called mesons and are very short-lived indeed. They disintegrate by stages until they eventually become electrons or other radiation.

The term 'meson' is rather misleading, for meson means 'between', i.e. occupying a state between the electron and the nucleon. Now certain mesons are, in fact, known to be heavier than protons.

The first mesons to be discovered were the pi- and mu-mesons. Since 1947, when cosmic rays were first produced artificially, scientists have discovered other types of meson. Particle accelerators can nowadays reproduce conditions in the upper atmosphere, and the scope of scientific observation has thus been greatly extended.

To give the reader some idea of what mesons are like we now present a brief description of the main types so far discovered. The reader would, however, do well to remember that there are *bound to be* other, still undiscovered, mesons.

The mu-meson

The mu-meson is a light meson, belonging to group L. It has a mass of 207 (the mass of the *electron* being taken as unity), and breaks up into one electron and probably two undetectable neutrinos (see p. 63). The entire disintegration is over in about two millionths of a second. Mu-mesons may be positive or negative, but, unlike other mesons, they have never been observed in an electrically neutral state.

The pi-meson

Group L also includes the pi-meson, one of the most common and widespread of all. The pi-meson is highly unstable and has a mean life of only 200 millionths of a second, after which it disintegrates into a mu-meson. It has a mass of 273. Unlike the mu-meson, it can exist in the neutral state, when it disintegrates, in about one millionth of a billionth of a second, into either one pair of electrons of opposite charge, two photons, or even two pairs of electrons. In the neutral state the pi-meson has a mass of only 264.

The K-meson

The K-meson is the bugbear of physicists. It has a mass of 966, and the mechanism of its disintegration is too complicated to be described here. At first it was thought that these mesons were part of systems, and names such as tau-, kappa-, xi-, and theta-mesons were given to each of them as it was identified in turn. Later it was discovered that all these mesons were simply different forms of one and the same type, which disintegrates in various ways, giving rise to three pi-mesons, or to a single mu-meson, or again directly to electrons. In all cases disintegration occurs within 100 millionths of a second.

The K-meson, too, is known to exist in the neutral state, but all its other characteristics have not yet been discovered, so that their full significance can only be guessed at. The more facts we discover, the more complicated does the picture tend to become.

The lambda-meson

A discussion of heavy mesons (i.e. mesons which are heavier than nucleons) introduces us to the problem of hyperons. Hyperons are very strange particles indeed, since, though the atomic nucleus consists exclusively of protons and neutrons, nuclear collisions give rise to particles heavier than the original nucleus. Clearly energy has suddenly been changed into mass. It is as if a player had changed a tennis ball into a football merely by striking it very hard.

We still have no idea by what mechanism these particles undergo such peculiar changes, all we do know is that lambda-mesons, and the two mesons which we shall discuss below, are hyperons, i.e. particles which are heavier than nucleons, and that, despite their very short lifetimes, they are universally present.

The lambda meson weighs 2,180 times as much as an electron, while the neutron from which it is ejected has a mass of only 1,840 electrons. Only the neutral lambda meson can disintegrate into either a proton and a negative pi-meson, or into a neutron and a neutral pi-meson. In other words this hyperon must be made up of a nucleon plus some other particle. The lambda meson disintegrates within three ten-billionths of a second.

The sigma-meson

The sigma-meson is found in the neutral, positive, and negative states. The neutral sigma-meson has a weight of 2,320. It disintegrates into a lambda-meson and a photon, thus generating a series of hyperons. The positive sigma-meson disintegrates in the same way, but produces either a proton and a neutral pi-meson, or a neutron and a positive pi-meson. The negative sigma-meson gives a neutron and a negative pi-meson. The mass of these last two mesons appears to be different, that of the positive being 2,327 and that of the negative 2,342.

The xi-meson

The heaviest meson of all is the ksi-meson, with a mass of 2,583. It gives rise to a neutral lambda and a negative pi-meson, and has an average lifetime of two ten-billionths of a second.

In addition to all the mesons we have discussed, there are other hyperons which, though much heavier, are more difficult to observe. Perhaps we may yet discover other mesons of groups K and L.

To sum up: after having deluded ourselves for far too long that we have a complete picture of nuclear processes, we now realize that the picture is far more intricate than we imagined. Scientists now suspect that microphysical processes still hide many a secret from us, and that there may well be even tinier micro-structures, as small in relation to nucleons as nucleons are to macroscopic phenomena.

ENERGY

Energy is yet another very strange aspect of space-time. It is not difficult for us to think of space, and of matter inside it. These two notions are static and easily understood. However, when we come to time and energy we are on shifting ground, for these notions are less concrete, and hence much more difficult to understand. Energy involves a dynamic factor—time—since energy is constantly being dissipated or accumulated. It may be emitted for a thousandth of a second, as in the explosion of an atom bomb, or for thousands of years, as in a geophysical catastrophe. In the first case the latent energy stored in the atomic nuclei is suddenly liberated; in the second case the earth's crust undergoes persistent compression, which culminates in periodic earthquakes.

These two examples give us some idea of two distinct ways in which energy can be accumulated and liberated. Transformations of energy are, in fact, the explanation of all evolutionary phenomena.

Atomic energy

To understand this point we must again return to the atom. We have seen that atoms combine into structures of varying complexity. While some molecules, such as those of water, which consist of two atoms of hydrogen and one of oxygen, are very simple, others, and particularly the so-called organic molecules, have a very complicated structure, each molecule containing tens of thousands of atoms. But in all cases it is the continual exchange of peripheral electrons which makes the atoms cohere. Thus the structure of all molecules depends on electrical phenomena.

61

The electron is held in orbit by the attraction of the central nucleus, and this attraction, too, is apparently of an electrical kind. Hence the chemical energy, on which our life depends, is largely derived from infinitesimally small effects, i.e. from intra-atomic magnetism.

The laws of electricity show that energy is generated whenever electrons are displaced within a system. The term 'electrical energy' covers a host of phenomena, all of which have a common factor: the electron and its electric charge.

There are, however, other types of energy, and nuclear energy is one of these. The forces which hold the nucleons of a given nucleus together are of a special type which has no counterpart. Irrefutable proof of the existence of these unknown forces has been adduced by atomic physicists, and the harnessing of nuclear energy (it is incorrect to speak of 'atomic' energy) is the direct result of man's ability to control these forces. Nuclear energy is in fact the most basic of all energies.

Atomic science has also brought to light other new types of energy, e.g. the energy produced by the interactions of those unstable particles which we have listed on page 57. When a pi-meson changes into a mu-meson, and the mu-meson in turn becomes an electron, energy is liberated. We know nothing about the real nature of this energy except that it is quite distinct from nuclear energy.

Undiscovered forms of energy

We have good reason to believe that our knowledge of nuclear energy is still very incomplete. Since 1930 research has been hampered by the failure of the law of the conservation of energy to explain certain facts. This law, which was once regarded as one of the cornerstones of physics, apparently breaks down when it comes to the nuclear emission of beta rays. The reason for this breakdown is that, as we have seen, nuclear energy is quite different from all other forms of energy. Scientists tried to solve the problem by introducing a new particle, the 'neutrino', but the neutrino is, in fact, a very weak manifestation of nuclear processes and cannot account for the vast energies liberated.

The neutrino is probably a neutral particle with no proper mass, which travels with the velocity of light. It may be an undetectable particle of pure energy, or it may be a particle of non-radiant energy.

Radiant energy, on the other hand, is more easily detectable. Thus light-waves carry a form of visible, radiant, energy comparable with invisible electromagnetic radiation. In passing from one orbit around the nucleus to another, the electron emits a series of waves, which behave very much like corpuscles and are known as *photons*. Photons resemble neutrinos in that they travel with the speed of light, have no proper mass, and are, in fact, energy in its purest form.

However, radiant energy is produced not only by electron transitions but also by the nucleus itself, since nuclear transitions, too, are basically of an electrical character.

Another type of energy is due to gravitational forces, and it, too, has characteristics that make it unlike any other. We know very little about gravitation, except that it acts on all matter at a distance. Albert Einstein based his general theory of relativity on the intimate link which may exist between gravitation and space-time. This link, which we have mentioned in connection with the Sputnik on page 45, is a most fascinating phenomenon. For while gravitation and mass are equivalent, and while nucleons have mass, they are quite unaffected by gravitation. Clearly this strange fact points to a hidden paradox.

Provisional summary

Let us recapitulate.

We have examined four types of energy which seem to be independent of each other. They are:
1. The energy which holds atoms together.
2. The energy which holds nucleons together.
3. The electromagnetic energy associated with light-photons. Chemical energy is no more than an indirect aspect of this type of energy.
4. Gravitational energy.

Curiously enough, life makes use of only a small part

of the total energy found in the universe, and, moreover, of energy in its less concentrated forms. In other words, the energy exchanges by which life is sustained are only a tiny fraction of the tremendous energies available. The stars, on the other hand, derive their fantastic energy from thermonuclear reactions. This nuclear energy is spread throughout surrounding space as radiant (and therefore electromagnetic) energy, to reach us after a series of very strange transformations.

Nuclear energy is not normally liberated on earth except under the impact of cosmic rays or through the natural disintegration of certain radio-active beta isotopes in the earth's crust. Gravity burdens us with weight and hence causes us to expend additional chemical energy. The tides are one direct and perceptible result of gravity. Most of the other effects are not directly observable.

Energy into mass

To say that one form of energy changes into another means a great deal more than one might think. In 1905, when Einstein established an equation relating energy to mass, he opened up vast new vistas. Since Hiroshima, the equation $E = mc^2$ is on even the layman's lips.

However, the equation does not imply that *all* mass can be turned into energy and vice versa. When the nucleus of an atom is split, part of its mass disappears; what was previously the energy fusing the nuclear particles

E   65

has become transformed into the kinetic energy (energy of motion) of the particles and fragments emitted by the nucleus. Now, this does not mean that *all* the matter in the nucleus has been completely annihilated or that any amount of mass can be created from energy. In fact Einstein's equation should be written:

$$\Delta E = c^2 \Delta m,$$

probably with discontinuous values of Δm. Here the Δ terms represent fixed quantities. To say that 'energy equals mass times the square of the speed of light' is meaningless. It is as if we were to say 'water equals hydrogen plus oxygen multiplied by some factor'. Einstein's formula merely expresses the fact that both matter and energy can be interchanged to some extent. The equation tells us how much energy is liberated when a certain quantity of mass is annihilated, and, conversely, how much mass can be created from a given quantity of energy. But the quantity of mass involved in both cases is decided by conditions of which we are still ignorant, and in any case these Einsteinian transformations apply to nuclear phenomena alone, and not to all matter.

Anti-particles

Here we must mention a fundamental fact: energy is changed into mass with the liberation of two particles, one

66

being the opposite of the other. Thus when a pair of electrons is formed out of gamma radiation, one of these electrons is negative and the other positive. Now the positive electron is called the anti-particle of the stable, negative, electron. Its lifetime is short, for as it collides with any of the abundant negative electrons of the atom, both are immediately destroyed. The two particles disappear to re-create the gamma radiation by which the positive electron was originally formed. The precise details of all these processes are still under investigation.

There are other anti-particles, as well. Much has been written about the discovery of anti-neutrons and anti-protons. These particles are produced during collisions of high-energy nucleons.

Ever since atomic scientists began to work with giant accelerators they have been able to bombard protons with powerful 'projectiles', similar to those found in cosmic rays, and to observe the results of the interactions between these projectiles and the nucleons of the target atoms.

However, laboratory techniques are still a long way behind nature's own methods. Cosmic rays contain primary particles, with an energy billions of times as great as that produced in the largest of our accelerators. Hence much remains to be discovered in this field.

Conclusion

By way of summary, we may say that the concept of energy is very difficult to grasp, and that much research

67

will have to be done before we can define it more accurately. We have drawn up a provisional list of apparently distinct forms of energy, though, in fact, it is extremely difficult to distinguish between them. Moreover, there may well be forms of energy, both on the cosmic and on the microscopic scale, which present-day instruments are unable to detect. There may also be special biological forms of energy, associated with living organisms.

Again, all forms of energy may have an unknown common factor. Einstein has tried to find this factor in his attempt to set up a unified theory, but even he came to realize that this task was too vast, at least for his time.

The fact that life itself is based on energy derived from chemical reactions is of the greatest importance. A closer investigation of this phenomenon may well give us the key to the mystery of life and suggest that it is part of a greater reality.

TIME

Time, that most mysterious aspect of our existence, is more difficult still to define. We say 'Time is a dynamic factor', or 'Time is an independent variable within the framework of space-time', or yet again 'Time is the thread connecting the beginning of the Universe to its end'.

But all these definitions are really quite meaningless,

since time cannot possibly be divorced from space. Thus the restricted theory of relativity, propounded by Einstein and Minkowski between 1905 and 1908, tells us that any point in the universe can be described only by four co-ordinates, three of space and one of time.

But time also has a more immediate meaning, which men can grasp intuitively (see page 105). It is likely that this meaning is impressed upon us by a periodic repetition of similar sense impressions. Our intuitive feeling that time is passing is largely based on such observable phenomena as the alternation of night and day, the seasons, etc. We cannot conceive of a universe in which nothing is ever the same.

Let us now look at physical time in all its varied aspects.

Astronomical time

Our unit of time is the second, roughly the time taken by one heart-beat, and by the swing of a pendulum of normal length (a pendulum of length one metre oscillates once every two seconds). The second is a unit invented by man for his own convenience, and has no objective existence.

Attempts have often been made to find a more objective unit of time, particularly of one that could be used for timing astronomical events. At first the alternation of day and night and the change of the seasons were thought to represent this objective standard, but they have since proved to be less reliable because they are affected by minor variations and irregularities in the earth's rotation. The earth is slowed down by the action of the tides, and gradually, but inexorably, our days are being lengthened, to wit by sixteen ten-thousandths of a second every hundred years. Hence, four billion years ago, when the earth came into existence, it must have completed a full rotation once every seven hours.

Another unit of time, used by astronomers, is the aeon, which represents 1,000 million years. Now when we say that the earth has existed for four billion years we must not imagine the 'year' to have been of equal length throughout the history of the earth. Originally the earth must have revolved at a much greater distance from the sun than it does now, and a year must therefore have been much longer. The age of the earth, as determined by radio-active disintegration of rocks, is therefore expressed not in 'real' years but in present-day years. Bearing this fact in mind, we may say that:

the Galaxy has existed for six aeons,

the sun for five aeons,

the earth for four aeons,

life on earth for two aeons,

the first known fossil for half an aeon,

mammals for about five hundredths of an aeon,

man for a thousandth of an aeon,

the earliest known civilizations for seven millionths of an aeon.

70

Time on the nuclear scale

A more reliable unit of time is based on radio-active disintegration. Nuclear stability is quite unaffected by such macroscopic phenomena as temperature, pressure, electricity, magnetism, and chemical changes. The nucleus is hermetically sealed against any contact with the outside world.

Now, an immense number of nuclei is in a state of continuous and regular disintegration, so much so that this disintegration can be used as an accurate yardstick for determining, say, the age of fossils 1,000–35,000 years old. These fossils are examined for traces of Carbon 14, which has a mean life of 5,600 years.

The nuclear unit of time is called the chronon. The chronon expresses the time taken for a light signal to travel from the centre of one nucleon to that of another, and each chronon represents 1×10^{-3} seconds, i.e. ten millionths of a billionth of a billionth of a second. Nuclear time can, in fact, be expressed on no larger scale. In comparison, macroscopic phenomena are almost eternal. Take, for instance, the nucleus of plutonium 239, which has recently been produced in an atomic pile, and which is one of a number of similar nuclei that go into the making of an atomic bomb. If the bomb is exploded after four months the nuclei will have existed for about a million seconds, but during each of these seconds ten million billion billion chronons will have passed. Translated into ordinary seconds, this interval represents three thousand million billion years!

Take a further example from recent experiments with elementary particles. It appears that the newly discovered mesons exist for between a millionth and a billionth of a second. While this strikes us as extremely brief, it represents an aeon on the nuclear scale. Measured against aeons, 'nuclear' time represents an incredibly small interval of the order of 10^{-32}

The secret of eternal youth

There is another, completely different aspect of time which must also be considered, viz. its inseparable link with space. According to the restricted theory of relativity, velocities approaching that of light lead to the appearance of an entirely new set of phenomena, phenomena that we are yet unable to describe. The greatest 'man-made' speed is that of the artificial satellite, which circles the earth with a maximum speed of about five miles per second. The earth, in turn, revolves about the sun with a speed of about eighteen miles per second, an insignificant figure when we compare it with the speed of light, viz. about 186,000 miles per second.

Now atomic particles, e.g. electrons, protons, and cosmic rays, can attain speeds approaching that of light, their speed depending on their energy.

As they approach that speed, the particles may be said

to have practically no lifetimes at all. Time has been cancelled for them. A light ray, i.e. a wave carrying a photon, which has exactly the same speed as light, is completely severed from time (according to our standards). While it leaves a star, one aeon away from us, to be observed on earth one billion years later, the photon itself 'experiences' its own departure and arrival as simultaneous events.

This fact must strike us as paradoxical, unless we abandon all our preconceived notions of time. The laws governing the behaviour of matter and energy apparently run counter to our common sense. Common sense holds good only on the macroscopic scale.

This 'suspension' or contraction of time can, in fact, be proved experimentally. Thus, mesons, produced by collisions of primary cosmic rays, manage to reach the earth despite their short lifetimes, simply because their velocity approaches that of light. In that way we can detect them, though on our scale they may be said to become extinct the moment they are created.

Neighbouring galaxies

There has been a great deal of discussion about the future possibility of putting relativistic physics to practical use. In a space-ship travelling almost as fast as light, time would become so contracted that the crew could cover vast distances without growing very much older. In that way men could travel to Sirius, nine light-years away, and return in about twenty years, but having aged by only a few weeks, months, or years, depending only on how nearly they travelled with the speed of light. This paradox becomes even more striking when we imagine journeys to stars as far as two million light-years away. Such journeys could be accomplished within the span of one human life, though, in terrestrial terms, the journey there and back would, in fact, have taken four million years. Hence men could reach Andromeda within a few years, to return to an earth that has aged by some four million years.

Space-time is also related to gravity. Thus any body placed in a gravitational field undergoes a time contraction in accordance with Einstein's generalized theory of relativity.

Now this slowing down of clocks is far too gradual to be detected except in intense gravitational fields like those of certain white dwarf stars which contain vast concentrations of dense matter. Even so, artificial satellites can now be used for observing this effect even for such relatively weak gravitational fields as that of the earth. Thus Sputnik II, which stayed up for twenty-four days, could have sent out time-signals from its two transmitters which, by comparison with two synchronized clocks on earth, serving as reference systems, ought to have revealed a time difference due to gravitational effects (the Sputnik was, of course, farther away from the earth's centre of gravity than the clocks).

.   .   .   .   .

While these aspects of time strike us as most peculiar, it seems likely that even stranger aspects may be revealed by future research.

Theorists have been discussing the reversibility of time for quite some while, and their work suggests that nuclear phenomena do not necessarily follow a one-directional time-scale. Thus anti-particles may well be particles which run counter to time, i.e. they may originate in the future and become extinct in the past. However, very much more research will have to be done before such theories are accepted. The course of time may be compared with a river flowing majestically between its banks. Are there whirlpools beneath the stream of time? Is there an ocean into which it flows and comes to rest? Is anti-matter destroyed at the time of our birth, only to be born at our own death? These are only a few of the questions which future investigators will have to answer.

4

The Contents: Life

DESPITE the vast number of facts he has at his finger-tips, the modern biologist still knows next to nothing about life itself, its origins, or its workings. Life exists in a host of forms; viruses, bacteria, plants, and animals are living structures, one and all.

The following section deals with some unsolved biological questions. It also attempts to point out some of the hidden treasures of nature.

It is only by analysing the miracles of nature, and by scrutinizing the results of our research, that we can ever hope to unravel the mysterious threads of our own existence.

Life, as we know it on earth, is subject to physical and chemical processes. Only if life were discovered on other planets could we hope to make a comparative study of the external factors that go into the making of us all.

We know very little about what life really is. What we do know is its structure, and it is this structure which we shall analyse first, beginning with its most elementary constituent, the molecule.

The study of biology usually begins with large organisms, then turns to smaller ones, and finally deals with microscopic creatures. Here we shall follow precisely the opposite procedure.

The chemistry of life

Everybody knows that nine-tenths of all living organisms consists of hydrogen, carbon, oxygen, and nitrogen. The preponderance of hydrogen and oxygen is probably due to the fact that, over two billion years ago, life on earth originated in water.

Among the vast number of chemical combinations found in living matter, some occur more frequently than others. It is important to know these, for the knowledge may lead us to an understanding of evolutionary processes. Chemical analyses of biological matter have shown the importance of certain substances known as 'proteins'. Proteins, which consist of huge molecules, make up the greater part of all living tissues.

Proteins are studied by modern chemistry, not only as a basic property of life but also for their general interest. The structure of each protein molecule contains several hundred thousand atoms arranged in long chains of amino-acids, of which more than twenty different types are known. Each amino-acid contains atoms of the four elements H, C, N, and O.

When two amino-acids lose a water molecule between them they produce a peptide, and gigantic chains of peptides go into the making of all proteins.

Spin

The inner structure of these enormous molecules has been studied in great detail. Long chains can perform a vast number of chemical functions, either by exchanging atoms or by adopting new ones. It is in this way that the incredible complexity of living matter has come about.

Proteins have a characteristic geometrical structure. Their molecules are arranged in spirals, and in all living proteins the spirals turn to the right. Hence all living matter is clearly asymmetrical. The reader may remember that the young Pasteur made a study of the symmetry of living matter in the crystalline state. Having observed the existence of a marked asymmetry, he began to wonder whether the fundamental asymmetry of matter was not a characteristic of life. Nowadays, physicists have discovered a similar asymmetry in nuclear phenomena.

Thus, the beta-particles emitted by radio-active nuclei are characterized not only by their mass and velocity but also by the direction of their 'spins', i.e. their rotation about their own axis.

Scientists have been investigating this problem, particularly in connection with radio-activity. In 1957, when experiments were made to test Lee and Yang's theoretical predictions, it was found that the emission of beta-particles was directly connected with the nature of the spin of both the nucleus and the beta-particles themselves.

We are concerned here with the practical results, rather than with the details of abstract, theoretical problems. The particles called neutrinos spin to the right, i.e. clockwise. Their anti-particles, the anti-neutrinos, on the other hand, spin to the left, i.e. in an anti-clockwise direction.

Anti-worlds defy us

It seems, then, that our Universe has a bias against the left, and that its symmetry is not nearly as complete as was previously believed. Now it is possible to imagine the opposite bias, i.e. a Universe with an 'anti-right' bias. Such a Universe would be the opposite of our own in every way. It would be an anti-Universe, consisting of anti-matter, made up of negative protons, positive electrons, neutrinos with a left spin, and proteins with left-turning spirals. The question then arises whether the Universe was not perhaps originally symmetrical, later to

79

be split apart by some catastrophe. In that case, half the stars and half the galaxies might well consist of anti-matter.

If there is life in these anti-worlds, do the huge protein molecules turn to the left? Is it correct to speak of an 'anti-life' as the mirror image of our own?

Such problems are extremely difficult to solve, though clearly their solution would shed new light on many basic questions.

At this point we must mention the peculiar nature of those macro-molecules which carry the hereditary characters of biological individuals. These molecules are found in the chromosomes of vegetable and animal cells, and are called desoxyribonucleic acid, or D.N.A. for short. Recently there has been a great deal of discussion about controlled mutations among ducks, using D.N.A. D.N.A. molecules are rather similar in structure to protein molecules. They contain chains of phosphates of sugars, purine, and pyrimidine residues. Now these chains, too, are spiral, or rather bi-spiral, for their two arms turn in opposite directions.

This important fact shows that the macro-molecules which transmit hereditary characters are fully symmetrical in contradistinction to all other living cells. This is a very remarkable phenomenon indeed. For while living matter, in so far as it serves to maintain the organisms' metabolism, is asymmetrical, genetic matter, which transmits the fundamental characters from one generation to the next, has a symmetrical structure. These facts may well lead to far-reaching conclusions, once the results of further experiments are known.

Viruses, links between man and the atom

When we consider organisms rather larger than molecules we come across fresh mysteries.

Viruses, which are made up of macro-molecules, are often confused with bacteria (microbes), for both were discovered in the same way. Pathogenic agents (i.e. agents causing diseases) were first identified by Pasteur. Now, since certain bacteria and viruses are pathogenic, and since the only difference between them was originally that bacteria could be seen through optical microscopes while viruses could not, size alone was thought to distinguish the two.

In the last twenty years electron microscopes have revealed the existence of viruses a hundred times smaller than bacteria, i.e. needing magnifications of from 50,000 to a 100,000 times before they can be studied. Bacteria, which can be seen under a magnification of 2,000 $\times$ are extremely complicated organisms by comparison. Viruses have all sorts of shapes. Some are elongated, some are spherical, and others again look like miniature tadpoles.

Two kinds of viruses must be distinguished, those which feed on vegetable matter and those which feed on animals. The former are smaller and less complex. They consist of several million atoms, organized into crystals which spread out into the infected vegetable tissue. Whether these crystals are really living substance is a question on which scientists continue to differ.

Animal viruses are rather larger and have a much more complicated structure, involving several hundred millions of atoms. They have not been found to exist in the crystalline state. Analysis shows them to consist mainly of D.N.A., unlike vegetable viruses, which consist mainly of R.N.A. (ribonucleic acid). Hence D.N.A., that primordial factor of life, makes its simplest appearance in animal viruses. No doubt this fact is of great significance, though the full implications still escape us.

Bacteriophages

Bacteriophages are animal viruses of a special type. They all have a 'head' and a 'tail', and like other viruses they have a hexagonal structure, i.e. the head has a hexagonal profile and the tail a six-sided section. The outer part of these viruses consists of proteins, and the inner part of amino-acids and D.N.A. When bacteriophages come into contact with a microbe they attach themselves to it by their tail and break its membrane by injecting it with D.N.A. from the head, using the tail as a 'needle'. In this way each bacteriophage generates hundreds of its kind. It takes about ten minutes for the bacteriophage to inject its own substance successfully.

Millions of different microbes

Even among the larger bacteria there are some that cannot be classified as living substances with any degree of certainty. Again, while millions of bacteria are beneficial to their hosts, others are pathogenic. The latter group includes those bacteria which cause such familiar diseases as diphtheria, plague, cholera, botulism, and recurring fevers. (Other diseases, such as measles, rabies, poliomyelitis, and smallpox, are due to the action of viruses.) The class of 'useful' bacteria includes the yeasts which cause fermentation, and those many forms which live in the soil and which are essential links in the nitrogen and sulphur cycles that enable plants to draw up nourishment.

A single gram of arable soil contains from six to eight million bacteria. An estimate of the total number of soil bacteria on the earth would involve figures of the order of a hundred billion billion billions. Very little is known about their complex internal structure. True, we know a great deal about their membranes, but even here much research will still have to be done.

The living cell

Protozoa are unicellular creatures which resemble the cells of more highly evolved animals. Protozoa exist in a host of different forms, though not nearly in as great a

83

variety as bacteria. More than 15,000 species of uni-cellular creatures have been classified; the amoeba is but one of countless examples. Spores, algae, foraminifera, and amoeba are all protozoa; hence the animal and vegetable kingdom are closely related.

Protozoa represent the simplest forms of life, for all other plants and animals are multi-cellular, i.e. they are made up of vast colonies of cells. Thus the human body contains some 500 billion cells. Groups of cells combine to make up specialized tissues, each with its own vital function and each working in harmony with all the others. Clearly, therefore, there is an immense gulf between single-celled protozoa and multi-cellular metazoa. As we go up the scale of life, we encounter atoms, macro-molecules, viruses, bacteria, and protozoa, in that order. Each of these is 100 times larger than the one preceding it—a mathematical connection that is highly suggestive.

LIFE: BETWEEN TWO INFINITIES

Our world is held between two extremes: the atom and the cosmos. Atomic and stellar processes alike are totally alien to our normal way of thinking, so that the notion of occupying a mere fragment between these two infinities is difficult to grasp.

Nature, however prolific, seems to be cast in a

certain mould. The size of the atoms and the very forces which govern their behaviour impose definite forms and precise dimensions on atomic assemblies. Molecules, for example, no matter how complex, are constructed according to a certain scale of fixed dimensions, and those compounds which, in their turn, are made up of molecules, obey the same rules. All physical phenomena must obey laws, the limits of which are, in the final analysis, determined by the nature of atoms.

Prodigies of life

These laws hold in every part of the Universe so far investigated. While there may be other phenomena which we cannot detect because they are not 'material' in our physical sense of the word, matter, energy, and light behave very much the same way on earth as they do in outer space.

These remarks apply, *a fortiori*, to the behaviour of living matter, which has adopted an incredible number of forms. There are hundreds of thousands of different living species inhabiting every region of the earth, from high up in the air right down to the depths of the ocean. When we think of the differences between an oyster and a mosquito, or a lobster and a pigeon, the spectacle of nature strikes us as varied in the extreme. It becomes even more remarkable when we remember that plants, too, are living organisms. Yet all these organisms obey the same basic laws.

Some examples of organized life

Life depends on material as well as on energy exchanges. The material foundation of life is the atom, which joins other atoms to form molecules, tissues, etc. The human body has 500 billion cells, each of which consists of some tens of millions of billions of atoms. Each living cell is extremely complicated. For example, the amoeba, though uni-cellular, is a living creature in its own right. In time it splits into two parts, and its life-cycle is begun anew, as it has been for at least 500 million years.

Other living creatures, e.g. microbes and viruses, are smaller still. Viruses represent the lowest form of life. We know that inert chemical substances can reproduce their kind, but unlike the amoeba, they develop inside a host, viz. a microbe, into which the 'parent' injects itself.

Life is a many-splendoured thing

These chemical structures represent the lower limit between life and inert matter. However, we must not allow this lowliness to obscure the fact that all viruses contain several billions of atoms, combined into very complex molecules.

The apparently immense difference between the virus's billions of atoms and the elephant's billions of billions of atoms becomes insignificant when we see them against the cosmic scale. To claim that the higher animals hold a position half-way between the atom and the cosmos is a sign of crass ignorance, for, in reality, we are much closer to the atoms than we are to the stars. We are earth-bound in more than one sense.

This does not mean that 'beings' a hundred miles in height or in length cannot exist on worlds larger than our own. What it probably does mean is that such creatures, if they do exist, would function quite differently from any form of life known on earth.

The sun is life

Clearly, our own life is the result of conditions which have existed on earth for hundreds of millions of years.

Our metabolism, i.e. those vital functions on which the absorption and expulsion of external matter depends, is the gradual result of certain given biological conditions, such as the composition of the earth's crust, of water, and the atmosphere. To look only at the last factor—if the earth's atmosphere consisted of, say, chlorine, and nothing else, no known form of life could exist in it. Organic molecules derive energy by oxidation, i.e. by combining with oxygen, and it is only because air and water contain an abundance of that element that life as we know it can flourish on earth.

Moreover, all animals must ultimately feed on vege-table matter. Plants alone can grow by synthesizing

87

substances from the air and from the soil. Their synthesis of complex organic molecules from simple substances depends on sunlight. Instead of losing energy by oxidation, as animals do, plants accumulate it. This fact is of great importance. A tree or a blade of grass accumulates energy; a man or a sparrow expends it.

Plants transform light energy from the sun into chemical energy by means of their green pigment, chlorophyll.

Hence, life on earth depends primarily on solar radiation. Thus when we burn wood all we do is to release the solar radiation stored by the original tree. Coal, which provides us with so much energy (in industry, heating, and transport), releases the energy stored by plants of the carboniferous age, 100 million years ago.

Food chains

Thus all animals rely on plant life for their existence. No animal (including man) can directly assimilate carbon, oxygen, nitrogen, hydrogen, and other elements (sulphur, phosphorus, iron, etc.) which are necessary to its existence, and must therefore obtain them, fully synthesized into complex molecules, from plants. Herbivores do this directly; carnivores do it indirectly, by absorbing what molecules they need from the flesh of herbivores. Fish, feed on other fish, which, in turn, feed on plankton, a mass of microscopic algae, each of which is a plant. Algae, too, derive their energy from solar radiation.

To sum up, while plants can synthesize organic chemical substances under the action of sunlight, animals must feed on plants. Solar radiation is therefore the cornerstone of all life.

Bearing this point in mind, it only remains to say that the nature of the synthesized chemical compounds depends on the nature of the soil and of the earth's atmosphere, and that solar energy is absorbed subject to the physical conditions of temperature, pressure, and gravitation found on the earth.

Our neighbour Mars

The basic conditions of life were once again re-examined when it was realized that plant-life probably exists on Mars. Seasonal changes in the colour of the dark areas (the Martian year is twice as long as our own) had long ago led scientists to suspect that plant-life might exist on that planet. They could not be sure, of course, for there might have been a completely different explanation for these colour changes. However, during the last close approach of Mars, in September 1956, astronomers were able to use new observational techniques. Their results were published at the end of 1957. From the solar spectrum reflected by Mars (or rather from its absorption bands), William M. Sinton was able to infer the nature of the planet's dark areas. This point requires amplification. Objects look green because all solar radiation except green (and such invisible radiations as infra-red and ultra-violet) has been absorbed on the way to us.

An examination of the sunlight reflected by the dark areas of Mars showed that certain infra-red bands were lacking, i.e. that they had been absorbed. The position of the bands indicated that the Martian molecules which caused the absorption contained carbon and hydrogen, i.e. the basic constituents of life on earth.

The tremendous implications of this discovery, which has been widely ignored, are plain for all to see.

An inhospitable world

Our own deep-rooted anthropocentrism has prevented our full appreciation of what the existence of life on Mars might mean. The physical conditions on Mars have been roughly determined by indirect means. The Martian atmosphere is very rarefied, with a surface pressure less than that found on Mount Everest. Most probably the Martian atmosphere contains oxygen, although in much smaller quantities than our own. The presence of oxygen is inferred from the red appearance of the planet, which shows that the soil must contain a great many iron oxides. These oxides are bound to have deprived the Martian atmosphere of much of its free oxygen. The force of gravity on Mars is only two-fifths of ours. Mars has little water, and the mean temperatures are decidedly lower than ours, although reaching 30 degrees Centigrade at the equator in summer. Mars receives a little less than half as much sunlight as we do, but as its atmosphere is very rarefied cosmic rays may be able to reach its surface much more frequently than they do our own.

If life has, in fact, developed under such circumstances, it must be adaptable in the extreme. True, human life, as we know it, can exist only within fairly narrow limits. Thus man cannot survive, naked and without shelter, at temperatures of less than 60°F. or of more than 140°F.—a very small range indeed. A caisson at a pressure of 5,000 atmospheres must be gradually decompressed for two hours if its personnel is not to suffer ill effects (the bends). At the summit of Mount Everest, on the other hand, the pressure is too low to provide our body with enough oxygen. If the ozone (O_3) in the earth's atmosphere did not intercept almost all the ultra-violet rays the whole surface of the earth would be completely sterile. Again, if our atmosphere were less dense, cosmic rays would reach us more freely and prevent the formation of any complex cells. But external factors affect different forms of life in different ways.

It must be clearly understood that, from its very beginnings, life as we know it has had to adapt itself to certain rigid conditions. These conditions are obviously not the same for all forms of life. Some microbes can live in paraffin; others are anaerobic, i.e. they do not require oxygen; mushrooms do not need sunlight; certain fish exist under pressures of 500 lb. per square inch; woodworms eat nothing but cellulose (a substance which we find completely indigestible); spores can survive at temperatures of below −300°F.; microbes can live in hot springs at temperatures of over 185°F. The problem of staying alive is solved by different forms of life in almost as many ways.

Life, then, is a general phenomenon, which may occur anywhere in the Universe, provided only that environmental conditions are not too severe. There is no doubt that forms of life can exist where human survival would be quite impossible and, moreover, in a host of unsuspected guises.

As men travel further and further into space they are bound to meet sights beyond their wildest expectations.

5

Perception

THIS section is the most important part of the book. Clearly the analysis of our means of contact with the external world is one of the prime tasks of science. What we know of our environment we know through our senses, or through their artificial extension by such technical aids as telescopes, microscopes, and amplifiers.

From our sense organs there has developed our brain, our consciousness, our intellect. The stages of this development are the stages of man's own evolution.

THE SENSES: DEFINITION

In what follows 'sense' refers to any system of organic stimulus-response mechanisms.

By our senses we come to know and to interpret the outside world, and since that world is very complex we need more than only one sense-organ.

But, quite apart from keeping us in touch with the

environment, our senses also inform us of our inner organic and emotional conditions. Thus our senses react during all external or internal changes, and transmit messages to the brain. Here the messages are not only recorded but also decoded, so that appropriate action may be taken. That action may be conscious or not, i.e. voluntary or reflex.

Our senses rule supreme

We have seen that our senses link us to our inert environment. In fact we might say that the senses are tantamount to life itself, since it would be impossible to imagine living organisms completely divorced from their surroundings. Any higher animal deprived of its sight, hearing, smell, taste, or touch would be unable to survive, if only because it could not avoid dangerous obstacles or obtain its food.

Even unicellular animals may be said to have senses. The amoeba, for instance, reacts to such exterior stimuli as light, chemical irritants, electricity, etc. True, its 'senses' are embryonic, but they are nevertheless present.

As we go up the scale of life we find that sense-organs become more and more elaborate. In other words, the higher an animal, the greater the range of its responses. With the introduction of the brain, nature interpolates consciousness between stimuli and responses.

Our sense impressions stimulate our thought to such

an extent that our consciousness may be said to be the direct result of the work of the senses.

Now since each organism has specific responses to specific surroundings, it is quite possible that, as environmental factors have changed in the course of evolution, some of our senses may have become blunted, at least in all but a few individuals. Again, since evolution is a continuous process, we may have latent senses that will become manifest in the future. What these senses are no one can tell.

THE FIVE MAIN SENSES

It is only natural to begin the description of our sensory equipment by a discussion of what are traditionally known as the 'five senses', all of which are associated with definite organs. Sight being the most important of the five, we shall discuss it first.

Sight

The eye is a highly complex organ. Its main task is to receive photons, i.e. those particles of light which were once thought to be vibrations of the ether, but which are now known to consist of waves of particles. These particles travel through space with a velocity of 299,792 km. per second.

Vast numbers of photons are produced during all

electron transitions in an atomic shell. Thus a candle will emit three billion billions of photons per second. Photons spread out evenly in all directions, and hence diminish in concentration with distance from the source, i.e. they obey what physicists call the inverse square law. This law enables us to calculate the number of photons reaching us from a distant star of average brightness. Each second, a star some 10,000 light-years away will emit a number of photons involving forty-five figures, of which our eyes will receive no more than four or five photons per second. When photons strike the retina lining the back of our eyes, they cause the decomposition of chemical substances and the consequent production of a fluctuating electric micro-current which is carried to the brain via the optic nerve.

A marvel of adaptability

Our eyes can stand the strain of billions of billions of waves of photons during the day, and yet respond to only a few photons at night. Moreover, the eye can distinguish photons of different vibratory energy (wave-length), i.e. it can differentiate between colours. Violet light has a wave-length of four ten-thousandths of a millimetre, while red is twice as long. All other colours of the spectrum have intermediate values. We saw in Chapter 2 that we are blind to all radiation beyond the ultra-violet or below the infra-red.

Many animals, however, can receive that radiation as well. Thus it has been shown that bees, unlike men, are sensitive to some ultra-violet rays and to polarization effects. By polarization we mean that the vibrations of the light-wave in reflected, scattered or transmitted light travel preferentially in a single direction or plane. Bees can therefore fly by the sun, since the intensity of polarized sunlight varies with the angle which the sun makes with the earth. Again, fish living at a depth of seven or eight hundred yards are particularly sensitive to infra-red rays, which penetrate further into water than other radiation. If we, too, could 'see' infra-red rays, i.e. heat, we should look upon our fellow men as glowing apparitions with particularly bright heads and hands, since these parts of the body are at a higher temperature than the rest.

As the eye passes all the light signals it receives through a lens, these signals are projected upside down on the retina. Newly born babies are therefore said to see everything bottom up. Later the brain learns to correlate these images with tactile impressions, i.e. it learns to reverse the picture once again.

Organs of vision can assume many different forms. Thus the eyes of certain birds are much more complex than ours, as are the eyes of deep-sea fish. Insects have eyes with independent facets, each of which can be likened to a complete eye.

Hearing

Hearing is almost as important a means of contact with the outside world as sight, though physiologists have not studied it to quite the same extent.

The ear picks up mechanical vibrations from the surrounding air. A compression in any part of the atmosphere sets up a wave which travels towards us with a velocity of about 180 miles per second, depending on the temperature. These sound-waves then set up sympathetic vibrations, first in our outer, and then in our inner ear. In the inner ear the vibrations are sifted according to frequency, i.e. the number of vibrations per second. The ear is sensitive to a range of from about ten to twenty thousand vibrations per second. Our musical scale covers an interval from one hundred to several thousand vibrations per second.

An acute sense

Like the eye, which can distinguish colours and intensities, the ear can distinguish frequencies and amplitudes. Just as 'pure' white light consists of a mixture of colours, so does 'pure' sound consist of a fundamental note and its complex harmonics. The tonal quality of an instrument is the direct result of this combination. The ear is also sensitive to variations in the volume of sound, from the softest rustling to the roaring of an aeroplane engine at close range.

The sense of hearing of animals is more highly developed still. A dog will respond to sound frequencies to which our ear is quite insensitive. Most probably human hearing, too, was more acute in the past, when conditions were more precarious and when man needed the fullest use of his senses to survive in a hostile environment.

Smell and taste

Smell and taste are very similar if not altogether identical. Our tongue can distinguish no more than four tastes: salt, sweet, bitter, and acid, so that the infinite variety of tastes known to us is really a result of our sense of smell. The smell of food in our mouth affects the organ of smell at the back of our nasal passage.

The sense of smell is perhaps more mysterious than any other sense. No theory can explain how a few molecules of perfume in a large room can, despite their scarcity, stimulate our olfactory nerve. All we know is that certain chemical substances diluted in air or in food make a characteristic impression on our senses, and hence on our brain. What is even more remarkable is that smells can be memorized. A certain scent may, by association, evoke events which took place many years ago and which have been completely forgotten. A dog's sense of smell is sharper and more developed than our own, and is often of far more importance to the animal than its sense of sight. A similar sensitivity to smell has been observed among insects. Thus certain male butterflies respond to the scent of their mates over distances of several miles.

Like all our senses, taste and smell keep us in contact with our environment. However, while these senses are essential to the survival of many animals, they have become blunted in man, who has learned to rely more fully on his other senses.

Touch

The sense of touch is unique in that each tactile nerve responds individually to pressure and heat. Degrees of pressure are clearly distinguished, for a blow from a fist and the prick of a pin produce characteristic sensations on the tactile nerve terminations of our skin.

On the other hand, our tactile organs cannot distinguish between extreme cold and extreme heat. They transmit identical messages to the brain, which alone can read them correctly.

Balance

Unlike most other animals, man stands upright. Hence the human body experiences an unusual gravitational pull, which demands a particularly highly developed sense of balance.

Our balance is controlled by an organ in the three semi-circular canals of our inner ear. These canals act as a sort of triple spirit level. Any movement of the head displaces a tiny particle in each canal, and the movement of these particles against the sensitive lining of the canals informs the brain of the position of our head.

To a large extent the sense of balance is supplemented by the sense of sight, by which we can tell directly whether we are standing upright. Nevertheless, the sense of balance can function quite independently, for we do not topple over whenever we close our eyes. Birds have a still better sense of balance, which enables them to fly steadily even when surrounded by clouds.

In brief, our organism is provided with a signalling system whereby external stimuli are transmitted to the brain, there to be decoded. The correct reading of the code ensures our survival.

THE WEAKER SENSES

In addition to the main senses which we have discussed in the last chapter, man also has a number of less definite, less obvious, and less localized senses, which I shall try to define and explain below.

Temperature

We are aware not only of external but also of our own body temperatures. We can tell whether we feel hot or cold, and we react accordingly. When we are too cold we shiver to produce heat by muscular exertion; when it is too hot we perspire and so lower the surface temperature of our skin by evaporation.

All warm-blooded animals respond to heat and cold. Their temperature sense, unlike their sense of touch, produces a reaction of the body as a whole, which is an essential survival mechanism. Since the metabolism of warm-blooded animals depends on the maintenance of a constant temperature throughout the body, and since our internal temperature is directly related to the external temperature, which varies continuously, we must have some automatic responding device.

Moreover, we are sensitive to changes in our internal temperature as well, particularly when we are feverish. This sensitivity is further proof of the presence of a special temperature sense.

Organ sense

There is another vague sense of which we are not directly aware, viz. our organ-sense. Thus we know vaguely whether we are in good, average, or poor health. It would need a whole book to discuss all the factors influencing our state of health. Mental and social conditions, climate and other environmental influences are only a few of the factors involved.

In what follows we shall do no more than list those

vague senses which keep us informed about our body's total relationship with its environment.

The hygroscopic sense

There is no doubt that the human body responds to the humidity of the atmosphere, if only because scar-tissue is especially sensitive to it. We feel uncomfortable in tropical damp or in mist and fog; a dry cold climate suits us very much better. Clearly, then, humidity affects us very strongly, and our hygroscopic sense plays a vital role in our lives.

The electrostatic sense

We are also very sensitive to sudden changes of the earth's electrical field or of its ionic composition. Storms cause great disturbances of electrostatic conditions, to which we respond, however feebly. Thus the old wives' tale that cats wash their ears just before a storm contains a measure of truth, since an abnormal amount of static electricity in the air makes a cat's fur bristle. The cat finds this sensation unpleasant, and tries to smooth its fur by stroking it with a moistened paw.

Modern urban life produces enormous quantities of static electricity, the effects of which on our health and attitude have not yet been sufficiently studied.

Vibrations and infra-sounds

There exists an immense range of mechanical vibrations, and although many of these may not be noticed by us, they must nevertheless have a marked effect on our responses. There are first of all those sounds which are too faint to be heard by our ear, since they are below our auditory threshold. Among these are the vibrations of everyday objects which are in precise resonance with the frequencies of common town noises. Thus factory machines and car engines running at different speeds cause sympathetic vibrations in their vicinity. For instance, window-panes vibrate with frequencies similar to, or identical with, those of Diesel engines.

A considerable proportion of these vibrations are infra-sounds, i.e. sounds below our own auditory range (which has a lower limit of twenty to thirty cycles per second). At the other end of the scale there are many ultra-sounds, with frequencies of more than 20,000 cycles per second.

Now there is no doubt that we react to infra-sounds even though we cannot hear them. Thus we are somehow aware of the slow swell of a heavy sea, of industrial background noises, and of a host of other inaudible sound effects.

Together with artificial electric effects, infra-sounds make the city-dweller's physical environment highly unnatural. They are certainly one of the factors responsible for the estrangement of urban man from those norms of healthy human behaviour which have prevailed for tens of thousands of years.

Direction

No one knows whether or not man has a real sense of direction. In this connection we must reiterate that a sense may have a very rudimentary form, either because it has not yet developed or because it has atrophied. Moreover, certain individuals have more highly developed senses than others, including a strong sense of direction, akin to that of many birds and other animals.

The sense of time

The last of the vague senses we shall discuss is an awareness of the passage of time. Discussions of this, more than any other sense, have always tended to be controversial.

There have been attempts to define a biological unit of time based on chemical processes inside the brain, and others to base the unit of time on cerebral pulses, heart-beat, and respiration.

There is a great deal to be said for such definitions, provided only that we are careful to distinguish between cause and effect. For the basic question is really whether 'time' has an objective reality or whether it is purely subjective—a reflection of man's perceptions and mental processes.

The answer to this question is that neither of these alternatives is correct by itself.

We shall illustrate this point by looking at the effects of fever and at the behaviour of animals whose body temperature is inconstant.

Fever speeds up our biological processes and hence our sense of time is changed accordingly. We 'live' faster in a state of fever and our impression is that time is passing very slowly compared with our bodily functions. If asked to tap his fingers every second, a feverish person will tap far more rapidly than he would normally do. He is inclined to cram more seconds into every minute, and so time strikes him as interminable.

A frog's life

Our second example is the frog, a cold-blooded animal, whose very life-cycle depends on external temperature conditions. A frog's egg will hatch into a tadpole in two days at a temperature of 90°F., in five days at 80°F., and in ten days at 60°F. Now the frog's entire life is influenced by the time which the egg took to hatch out. If a frog's egg is spawned and hatched in warm water the frog which develops from it will grow and live at a rapid rate. A frog originating from an identical egg spawned at the same time, but hatched out in cold water, will grow much more slowly, and by the time the first frog is old, the second will still be young. Here we have a striking resemblance to the space-travellers of the future, who would return to earth as young men long after their children had died of old age.

If we were frogs, we should have to have a radically different method of calculating time. Warm days would be very long and cool nights very short. Some warm-blooded hibernating animals, such as the brown bear or the ground-hog, must have a similar impression of time. During their long sleep their metabolism slows down so much that time must pass them by very quickly indeed.

These examples give us some of the clues needed for solving the mystery of our time-sense. Our perception of the passage of time is largely subjective, but it also appears to be based on an objective factor that is much more difficult to observe and to analyse. A great deal of research into the lives of animals of all orders will be needed before we can settle these aspects of 'time' unequivocally.

AUXILIARY SENSES

While slowly evolving towards a better understanding of his environment, man has had to improve his natural faculties and even to invent new 'senses'.

To put it more precisely, scientists have widened the range of our senses by discovering means of extending them, i.e. by inventing instruments which can reveal normally hidden phenomena.

It may therefore be said that science has endowed man with new senses by providing him with new tools. It is to these tools that we have applied the name 'auxiliary senses'.

Looking at atoms

For instance, the electron microscope, which has stepped into the shoes of the optical microscope, enables us to *see* what was previously quite invisible.

The demonstration of such small-scale phenomena is one of the most remarkable achievements of the human intellect. The most powerful electron microscope (the Berlin Elmiskop) can distinguish details down to a diameter of seven or eight Ångström units (the Ångström represents a hundred millionth of a centimetre). An atom has a diameter ranging from one to two Ångström, according to its complexity, and simple molecules consisting of ten atoms have a diameter of from six to eight Ångström.

Hence we are on the point of actually seeing all sorts of atoms, and Tungsten atoms have, in fact, been photographed. In other words, the basic constituents of life can now be measured and analysed.

The electron microscope, a fantastic extension of the human eye, has advanced our knowledge of visual phenomena beyond all previous expectation. Even more, however, will be achieved, once it has been developed to the point where we can use it to investigate sub-atomic particles. For the last thirty years scientists have been able to follow the trajectory of electrons and other charged particles, either in cloud chambers, on specially sensitized photographic plates, or, more recently, in bubble chambers. However, all they can see in these instruments is the *track* of particles as they cause condensation of the surrounding water droplets, etc., by ionization. Cloud chambers have been used to study atomic disintegration (lifetimes) and the relationship between various charged particles. (Neutral particles, which fail to ionize gases, cannot be observed in cloud chambers.)

Looking at the stars

At the other end of the scale we have those giant structures which make up the stellar universe. The fact that we can photograph stars a billion light-years away, and analyse their spectra, represents yet another tremendous advance in scientific techniques. Here again the power of the eye has been increased several million times. There is surely no finer illustration of human grandeur than the way in which man, a tiny prisoner on a tiny planet, has been able to look across vast cosmic distances and to analyse messages which have travelled through space for aeons.

Even so, many phenomena will always remain invisible to us, for instance, most electro-magnetic vibrations, of which visible light represents no more than a tiny fraction. The Universe is filled with photons representing a host of wave-lengths. Thus the interval between gamma rays emitted by atomic nuclei and infra-red rays covers a total of forty octaves. Moreover, a wave-band of fifteen octaves is now used to transmit radio and television signals covering a range of from ten kilometres down to one metre. Electro-magnetic waves alone cover a total interval of from a hundred billionth of a centimetre to ten kilometres, representing a total of fifty-seven octaves. Of these only a single octave (from 0·4 to 0·8 thousandths of a millimetre) is visible to the eye.

The eye is blind

When we consider that fact we realize how weak our organ of sight really is.

So far we have discussed electro-magnetic (photonic) radiation, but there is also corpuscular radiation which we shall be discussing more fully on page 121. For the moment we shall merely state that scientists have developed instruments to detect the particles involved in this radiation, and to measure their charges, masses, lifetimes, and velocities. (The velocities involved are generally close to the speed of light or else simple fractions of it.)

Infra and ultra-sounds, i.e. sounds we cannot hear, can also be detected by instruments. The human ear has a range of ten octaves, ranging from twenty to 20,000 vibrations per second, and there are many octaves which we cannot hear. Here we have a further example of the way in which man is adapted to his immediate environment. To survive he has no real *need* to listen to the electro-magnetic vibrations of a remote universe. All he does need to hear are the vibrations of the atmosphere in which he lives.

The electro-magnetic sense

Considering the importance of electrical phenomena, it is surprising that man has never developed a real electro-magnetic sense. True, we are aware of the degree of static electricity in the air, but we have no organ for responding to the motion of electric charges. Yet there is no doubt that life is basically dependent on electric flow, since our nerve-impulses, and our brain activities, involve the displacement of charged particles. Now any displacement of electric charges sets up magnetic fields, however weak and undetectable.

Lacking the natural means of perceiving these fields (unlike certain fish, which can emit and detect electrical impulses), man has had to provide himself with an artificial electrical sense.

Electrocardiographs and electroencephalographs now measure the electric potential of the human heart and brain ('brain-waves') very accurately.

Future senses

We have seen how science has extended the range and scope of our senses. Even so, many of nature's mysteries continue to elude us, not only in the environment but in ourselves as well.

While this ignorance persists man will for ever search for means to dispel it, even though all that new knowledge must fit his limited power of comprehension. This point is best illustrated by the example of that elusive particle the neutrino, which apparently plays an essential part in certain nuclear processes. It is a purely theoretical particle, with undetectable properties, since it has neither charge nor mass, and since its material effects are so negligible that it could travel across the sun a billion times without once colliding with another particle. Nevertheless, if neutrinos really do exist, the universe must be teeming with them. The sun must emit vast quantities of neutrinos and so must our own atomic piles which have been in existence since 1942. Now, since the neutrino escapes all our methods of investigation, we have been reduced to representing it in purely mathematical terms. Yet who can tell whether the future might not show us the way towards 'sensing' this and many other elusive phenomena?

The human brain functions in an extremely complex way, so much so that physiologists still know very little about how the brain interprets messages, reflects on these messages, and makes decisions.

In this section only two of the many functions of the human brain will be discussed which involve a veritable sense, *viz.* the musical and the mathematical. These senses seem, in some way, to be related to each other.

The brain contains the greatest sense

Mathematical arguments and mathematical intuition may reasonably be said to spring from a mathematical 'sense', for they are means of analysing reality. However, our mathematical sense must be distinguished from all our other senses, all of which discern material phenomena. The only material factor governing the mathematical sense is the practical experience which underlies all mathematical thought. But as mathematics advances, it casts off the shoes of empiricism to step into pure abstraction. Ever since Evariste Galois made his brilliant contribution to mathematics, mathematical thought has explored a world of ideas so far removed from experience as to correspond with no known reality. The great mathematicians of our time can be said to work by intuition rather than by external sense data.

By means of his purely cerebral mathematical sense man has made discoveries that could not possibly have been made by any physical means. True, when a mathematician postulates the properties of generalized space he starts out by considering existing physical space, but his results far transcend his original premises.

We may well ask if 'existing' has any real meaning in this connection, for all 'existence' may well be based on subjective reactions, i.e. it may be impressed on reality by our brain. An entity which 'exists' in five dimensions of space plus two dimensions of time may conceivably form a mathematical 'idea' of a four-dimensional Universe such as ours, though its own Universe would be altogether different. But its mathematical speculations about our world would in no way entitle it to postulate its real 'existence'. To it, our world would always remain a pure abstraction.

The spider: an Einstein in its own way

Is man the only animal endowed with mathematical ability? He is certainly the only one to formulate purely abstract mathematical ideas, though he is not alone in his mastery of geometry and topology. Spiders act as if they had the brains of first-class mathematicians. For them, space has properties which make its structure quite unique. Moreover, these properties are dynamic. Each time a spider spins a web it must cope with changing conditions, for the construction of a web involves such variable factors as wind direction, weather protection, exposure to sunlight, and the abundance of prey. The web itself is a masterpiece of construction. It has all the ideal properties which engineers look for—maximum resistance and maximum efficiency, combined with minimum use of material.

The silent bee

The bee is another insect which seems to have great mathematical gifts. Honeycombs are built according to maximum efficiency principles. Being hexagonal, the cells make use of available space in the most economic and symmetrical way possible, and the angle between adjoining cells is such that the smallest possible amount of wax is required for their construction. Moreover, bees inform one another of the presence of food or of good swarming conditions by performing geometric dances. Perhaps the most striking of the bee's mathematical activities is its ability to count. When bees swarm, the whole population of the hive, numbering several thousands, is divided almost exactly in two.

I have mentioned spiders and bees in order to show that mathematical thought is not restricted to man alone. Nevertheless, our own mathematical thought is far and away beyond anything found in the rest of the animal kingdom.

The music of the higher minds

Our musical sense is closely related to the mathematical. Almost invariably, great mathematicians have a deep love for music, though there is no evidence that the reverse is equally true. Great musicians do not necessarily have a profound understanding of mathematics.

It would require a separate volume to deal adequately with the musical sense. Like the mathematical, it is purely cerebral, i.e. purely subjective, hard to communicate, and highly personal.

From our point of view the most interesting problem is not that of musical 'appreciation', for the way in which music communicates its message to the listener can be explained in terms of rhythm and harmony. The real problem is that of musical 'creation'. Great composers seem to have an abnormal understanding of a world that is completely closed to others. Their highly developed musical sense allows them to feel and to express what no words can describe.

Those who listen to music are transplanted into a world that cannot be reached in any other way. Thus music seems to evoke responses from a sense that is still embryonic in most of us, though present in all. The musical sense may be compared to the sense of love, another purely mental state.

Our senses not only serve as links between us and our inert environment they also relate us to our fellow men. In non-humans the collective sense is usually attributed to 'instinct', an altogether meaningless term. 'Instinct' explains nothing, and is no more than an excuse for not thinking about the problems involved.

Productive associations

Many animals have a group-sense, by which they can live in harmony with their congeners. Community life offers each individual certain advantages and increases his chances of survival. A good illustration is the behaviour of emperor penguins, which has been studied by French polar expeditions. These penguins congregate at the beginning of the Antarctic winter and live in colonies for several months. During that time the eggs are hatched and the young birds reared until they can follow the adults to the open sea. While they are still helpless the whole community combines to protect them. Thus during severe blizzards the adult birds shelter the young by forming a living wall around them, each penguin pressing closely against its neighbour. The 'wall' is thickest where it faces the icy wind, and each penguin takes its turn on the outside of the wall, exposed to the full force of the blast.

There are other species of penguin which display

even more extraordinary communal traits. Shortly before the eggs are laid these penguins build nests of pebbles. Pairs of birds roll the pebbles slowly and laboriously to the nesting site. Occasionally a penguin will take a pebble from a stranger's nest, and should the theft be discovered the whole colony will punish the culprit by pecking it or beating it with blows of the wings. On the other hand, should any member of the colony suffer an injury preventing it from travelling with the flock, two of the colony will stay behind to tend it until it has recovered.

We could give many other examples of the behaviour of gregarious animals, for instance of such insects as bees, ants, and termites. These are, however, too well known to require further discussion.

Attuned individuals

Man's strongly developed social sense may well have an organic basis. There have been many theories about sub-vocal communication by special 'waves', though there is little evidence to prove that such waves really occur. At the same time we cannot reject that possibility out of hand. The existence of something in the nature of a vibration, wave, or some other physical phenomenon cannot be denied on the grounds that it has not been detected.

The very existence of society makes gregarious behaviour a biological necessity. Such behaviour may well be comparable to the phenomenon of 'resonance' in physics. Resonance plays an important part in all mechanical and electrical vibrations. Every material object and every electrical current has a specific vibratory frequency, i.e. the frequency which it would normally emit during primary oscillations. Any other primary oscillator in the vicinity causes the first oscillator to vibrate in sympathy. For example, a window vibrates in resonance with the vibrations of the engine of a passing car. Sometimes the air vibrates very strongly when planes fly past at very high altitudes. Wireless sets receive signals because their circuits are tuned to the same electro-magnetic frequency as the transmitter.

Crowds as resonators

Now, men congregate in response to a phenomenon very similar to resonance. Indirect proof of this contention is found in our crowd behaviour. This is a well-known physiological fact. In a crowd man ceases to be an individual. He loses some of his self-control, no longer reasons clearly, and is swayed by mass hysteria, so much so that he begins to think and act like the rest of the crowd. We have only to think of football matches, political or religious get-togethers, or even of crowded streets. When several hundred people congregate a new sense intervenes. If the individual feels at one with his neighbours their feelings become attuned. On the other hand, if the crowd lacks a general purpose the individual begins to feel instantly isolated. Some people sense this isolation very strongly. They suffer from agoraphobia, and the presence of even a small group of people makes them ill—their security is threatened by all strangers.

The most wonderful sense of all

There must be a similar explanation for our sense of love. Love has two aspects. In one sense it is purely biological and based on sexual attraction. But love also assumes another, much more cerebral and deeper form. Quite possibly, resonance is part of the reason why two people can be so much at one that they share the same feelings and ideas.

How is it that two people can be irresistibly attracted to each other at first sight? Surely it is because they are 'in tune' with each other, in the sense that their two systems vibrate with the same frequencies. Now there is good evidence that we all emit certain 'waves', which obey the laws governing all vibrations. The waves explain why some people will produce 'instinctive' loathing in some of their acquaintance, while others find them pleasant enough companions. Similarly there are many couples in whom love is completely one-sided. One person feels in harmony with the waves emitted by the other, but his own waves are rejected by the object of his devotion.

Waves are perhaps the best explanation of many of

our feelings. The sense of love, the sense of sympathy (or antipathy), crowd hysteria, and others may all be different aspects of one and the same sense. Insect communities and human societies are comparable in as much as their collective sense originates in the fundamental structure of all living organisms.

We lack organic geiger-counters

Man, though superior to other animals, is far from perfect. His senses are 'closed' to many marginal phenomena (ultra-violet rays, ultra-sounds, faint smells, etc.), and he is completely 'blind' to nuclear radiation or cosmic rays. In other words, we have no organ for responding to the impact of elementary particles which bombard us all the time.

These particles have such great energy that they can penetrate matter itself; they may be protons, neutrons, electrons, or mesons. While they do not affect us to any great extent, small traces of them reach us from outer space in the form of cosmic rays, or from uranium and thorium disintegration in the earth's crust. Radio-activity due to these elements can be detected in sub-terranean springs and in the atmosphere which contains minute quantities of radon (atomic number 86). All living organisms contain traces of radio-active potassium, for one of the isotopes of this vital element is naturally radio-active, and has a lifetime of several billion years.

Still, living organisms are exposed to very little particle radiation. Assuming that man's eyes were adapted to detecting this radiation, this is what we should 'see': The sky would look somewhat darker over countries situated between the equator and a certain latitude; above this latitude it would be brighter, for cosmic rays hit the earth only within certain latitudes, depending on the strength of the earth's magnetic field.

We should 'see' kerbs as being much brighter than the rest of a street, as kerbs are made of granite which contains a higher proportion of the radio-active elements thorium and uranium. Some regions of the world, such as Brazil and the Indies, would look much more 'luminous' than others, for their soil is much richer in radio-active isotopes.

Most noticeable of all would be the luminosity of the air we breathe and the food we eat, due to the accumulation of radio-active particles from the experimental nuclear explosions of the last ten years. The clouds overhead, too, would be extremely bright, and produce brilliant showers, returning these radio-active particles to earth. Nuclear explosions themselves would create such an immense concentration of radio-activity that the effects would be blinding.

If our organs were sensitive to radio-activity we should feel greatly irritated by radio-active nuclides, traces of which we breathe in and swallow with our food to store them particularly in our bones.

From the artful crab to the magnetic pigeon

There is no reason to assume that all animals are equally insensitive to radio-activity. For instance, colour variations in certain shallow-water crabs may well be caused by their special sensitivity to cosmic rays, since these changes do not result from changes in light intensity, nor from any other known periodic fluctuations. Since the changes vary with depth, it is quite possible that they may be caused by the progressive absorption of cosmic rays by sea water. This theory is, of course, highly speculative.

Another phenomenon to which we are 'blind' is terrestrial magnetism. While its effects are rather weak, it is odd that nature has not provided us with some organ to detect it, since such an organ would serve us as an excellent guide. The marvellous sense of direction found in migratory birds has often been attributed to a special sensitivity to magnetism. These birds are said to be aware of their position, or at least of their general direction, by being able to tell the magnetic intensity of any spot on earth. This explanation is borne out by the way in which homing-pigeons become frightened and confused whenever they fly past a radio-transmitter emitting electro-magnetic waves of much greater intensity than that of the earth's magnetic field.

Age-old radar

As we have seen, we are not only blind to many rays but also deaf to many sounds. In that respect we are not nearly so well equipped as bats, which use ultra-sounds as we use radar. Bats emit ultra-sounds, which rebound from obstacles and tell the bats how to avoid collisions. At night a bird may fly against a lighted window, but never a bat. For although the window is transparent to light it is opaque to ultra-sound.

It is now known that some fish, e.g. eels, emit electric impulses. These impulses have an even more striking resemblance to radar than those produced by bats, though they are electrical rather than electro-magnetic radar oscillations. It is reasonable to suppose that these fish can receive as well as emit the impulses, and that they can act accordingly. This is another sense which we lack completely.

We end this list of our missing senses with man's blindness to very high frequency photon radiation. If we were sensitive to wave-lengths from one metre to one centimetre, the sun would appear to be twice its normal size, because we should then be able to see the gaseous layers of the sun's atmosphere. Our idea of the size and shape of the sun is very subjective, for our eyes can see only that relatively small region which emits visible photons. Again, if we had an organ for responding to gamma rays we should 'see' a small sphere in the centre of the sun, i.e. the region where thermonuclear reactions occur.

6

The Great Future

PHYSICISTS are busily trying to synthesize the findings discussed in the previous sections, aware that our present ideas are but a fragment of what future generations will come to know.

Man's place in space and time will never be explained more fully unless science undergoes a revolutionary change of heart. Only then will our feeble attempts be rewarded by a truly magnificent vista of man's immense potentialities.

SCIENTIFIC THEORIES

To end our survey we shall examine briefly the future of the human race as a whole.

As time passes, the human mind grows ever keener and more adept at discovering empirical facts and the laws they obey. We cannot say whether this process is a natural part of evolution, and therefore permanent, or merely a temporary phenomenon that may not continue for more than a few thousand years. It is possible that our present civilization is only a stage in an evolutionary pattern that alternates between progress and reaction.

The important thing, however, is to remember that within a few centuries the human mind has come to control so many essential aspects of its environment.

Undoubtedly the most important factor in acquiring scientific knowledge has been rational research based on experiment. We are using the word 'scientific' in its widest sense, to include not only the many specialized branches of physics (in the narrow sense of the word) but also chemistry, biology, sociology, and psychology. Most of these disciplines involve the use of specialized mathematics.

Mathematics is not a science in its own right but a way of thinking about a host of topics. Mathematical studies reflect stages in our intellectual development, and hence in our respective preoccupations. The future may well produce new methods of thought far outstripping the physico-mathematical tools of the modern scientist.

Even so, there is no doubt that physico-mathematical theories have played a major part in scientific research during the last 200 years. Experiments have produced a plethora of new data, which have been sifted, classified, and interpreted, however crudely. Experimental research is continuously filling the gaps in our knowledge. Many discoveries still remain to be made, if only because our new instruments have not yet been used over long enough periods. Now, whenever experiments are repeated over long periods certain discrepancies are observed which cannot simply be attributed to 'chance', or to experimental error, and which must therefore be ascribed to basic theoretical misconceptions. After all, modern theories are the results of relatively recent, and, therefore, relatively short-lived, experiments.

If we consider the sum total of man's knowledge, and the instruments by which he arrived at it, we must conclude that the human mind oversimplifies to an unwarranted extent. But then it is the function of mathematical theories to reduce the vast number of dissimilar phenomena, which are the world as we know it, to the most logical and concise form possible. The atomic theory of matter is an excellent example of how an almost incomprehensible mass of data can be correlated.

Simplification, however, is often carried to ridiculous extremes. The belief that all phenomena can be made to fit a single pattern is rife in every school of theoretical physics throughout the world.

Yet it strikes me that the faith in a Universe which is simple enough to be defined by a few equations is fundamentally unsound. Simplification has played a very important role in all our physical theories over the past fifty years, but we must remember that all these theories are based on a restricted number of phenomena, observed over a limited period of time. The current faith in the validity of theories such as Einstein's and Heisenberg's, which try to unify all the known physical laws may be perfectly warranted, but to my mind these theories are a stage in human thought rather than its culmination. In its present state theoretical physics is rather like a pedlar's coat. The coat is shapeless and flapping, sadly patched up with a thousand clumsy stitches. Our scientific tailors hope that some day they will be able to make a coat which will fit us perfectly, and which will be absolutely up to date. While theirs is a noble ambition, it must be remembered that even the most fashionable of coats may become threadbare tomorrow.

It is quite possible that human thought itself develops according to succeeding 'fashions', and that as science advances, radically new modes of scientific thought will emerge and lead to quite unsuspected discoveries. Future generations will look upon the ideas that we consider so final today as mere steps in a historical process.

Faith in the underlying simplicity of the cosmos has been proclaimed time and again. Personally, I do not share this faith, nor do I believe that nature can ever be described by a few symbols. It is man himself who simplifies everything that is too complex to be grasped by simple thought, for man is inherently incapable of dealing with more than one thought at a time.

It is possible that advances in knowledge follow a cyclical pattern: experiments—theories—theories—experiments. Data are collected from the initial experiments and interpreted by theories. These theories give rise to others, which suggest further experiments. The results of the new experiments show the inadequacy of the theories on which they were based, new theories are propounded, and so *ad infinitum*.

In short, it is most arrogant to ascribe to the human mind an unlimited capacity of understanding the immense complexities of the Universe. Clearly, scientists of the future, too, will continue to waver between certainty and doubt, for despite the accumulated knowledge of generations, their knowledge will be no more than a smooth pebble from the shore of the great ocean of truth.

THE CONSCIOUS UNIVERSE

Is man the most highly developed organism in the whole of the Galaxy? Is human intelligence unique?

While it is impossible to answer these questions objectively, we know that man, by virtue of his intellect, is in a class far above any other *terrestrial* species.

Just the same, when we consider that the earth is no more than a tiny fragment of the cosmos it is only natural to wonder if creatures comparable to man exist on other planets. To answer this question we must first take several factors into account.

What we know about the many planets on which life could exist in other galaxies, and particularly in our own, leads us to think that life as such may not be restricted to the earth. The fact that vegetation exists on Mars proves readily that there are billions of potential life-supporting planets in the universe.

There are two possible ways in which life in other worlds may be organized. It may be so unlike any form we know that we should be unable to communicate with it, since, after all, communication implies similarity of psychological structure. And even if we could contact such organisms in some way they might be unable to respond intelligibly. For example, they might 'speak' by gestures a hundred times more complex than the dances by which bees convey information to one another.

The second alternative is that the higher forms of life on other planets resemble our own. This is a definite possibility, for even though the necessary conditions for the evolution of such forms of life are extremely rare, the Universe is so vast and so varied that they are likely to prevail in some parts of it.

If they exist, have these humanoids reached a stage of evolution comparable with ours or are they beyond or below it? There is no real reason why life in the Universe should not be found in all three stages. A million years has little significance on the galactic time-scale. The Universe might contain as many different degrees of civilizations as there are habitable planets. Moreover, there is no reason why any given civilization should not be in three stages of development simultaneously, for the notion of simultaneity need not apply in other parts of space-time. We ourselves have evolved according to our own time-scale, but other worlds are so far away that we cannot apply the same standards to them arbitrarily.

The fact that beings resembling man, though with higher technical accomplishments, have never visited us, is sometimes presented as 'proof' that they do not exist. However, there is no real evidence that they have not, in fact, visited us in the past. The earth has existed for four billion years, and man has lived on it for a million. Of these million years only four thousand have been partly recorded as human history. Surely there have been ample chances for beings from outer space to visit us unobserved.

On the other hand, no such visits need have occurred, even if space-travel were within the means of 'men' from outer space. We dream that we shall be able to travel freely through space some day. Yet we must remember that, at best, we could reach no more than a few worlds, since millions of years would be needed to visit some on which life might exist. This may be one reason why the earth has not been invaded and colonized by creatures from other planets, for their own journeys would take ages, and there are, moreover, millions of other planets which they could explore.

Then again, man may equally well be the first and

only creature to have acquired the technical and intellectual powers he has. Perhaps life and consciousness are phenomena which have found their cosmic climax in man alone. That, too, is a definite possibility.

It seems that man is firmly anchored to his small sphere of rock and soil by thousands of subtle links, and that he is fated to commit many errors, to the great disadvantage of himself and his innocent heirs. It has been suggested that some novae, those stars which suddenly explode, may have been blown up by their inhabitants who, like us, have juggled with cosmic and atomic forces. This suggestion was, of course, not made in full seriousness, but we would do well to bear it in mind.

MAN'S MORAL SENSE AND SCIENTIFIC FUTURE

The driving force behind man's evolution has never been fully explained. Mankind lives on top of a seething cauldron below the earth's crust. The end-results of this conflagration are anyone's guess.

While man's past achievements have not always been for the best, the gradual evolution of human intelligence reads like an epic. That epic continues to be written, and all of us are its *dramatis personae* capable of making however small a contribution. Mankind is an aggregate of billions of character traits inherited from our ancestors. Our own heirs, in turn, will inherit *their* traits from us.

We really know nothing of man's destiny, though man's ability to fashion his future, one of the factors of human greatness, may give us a clue to what lies ahead.

Every technical achievement has always been a victory over the future. When man first began to invent, his main object was to guard against future dangers or to make sure of food supplies. Primitive discoveries always had the future in mind.

No wonder then that those colonies of insects which seem to us most highly developed (we might almost say 'most civilized'), are those which apparently act with forethought. Termites, ants, and bees all work to ensure the survival of their species by creating suitable conditions well in advance. Similarly, beavers construct dams against future contingencies.

Man's awareness of the future plays a large part in his conscious strivings. These strivings, again, are closely related to his moral sense, for thought for the future implies mutual aid, and hence neighbourly love. Morality is the code whereby man achieves common goals with his fellows.

Seen in this light, civilization takes on a wider meaning: the conscious evolution towards interpersonal freedom and respect, and the creation of new values.

The two beacons which guide man along his difficult path are therefore morality and forethought. If one or the other of these beacons should begin to grow dim, disaster would strike at once. Knowledge, and hence a science based on true morality and real forethought, are invaluable aids to keeping our lights aflame. Any other sort of science is not only useless but positively dangerous.

Let us look more closely at the close connection between forethought and morality. So far, technical advances, basic discoveries, and new methods of thought have combined into a single approach which we may call science. Through science man has gained great, though incomplete, knowledge and control of himself, his immediate environment, and the Universe.

This power of controlling the environment might well lead towards a greater and better life for us all, but only if it is guided by a collective moral sense.

Thus we have reached a critical point of history. Unless we think very deeply about the consequences of recent discoveries, and do something to control nuclear experiments, we run the gravest risks of total destruction.

Any scientist worth his salt is bound to smile rather bitterly at such phrases as 'science for science's sake', 'pure knowledge', or 'ivory-tower isolation'. The atomic bomb has blasted not only Hiroshima but the entire house of cards that nineteenth-century scientists tried to erect as a monument to their own glory. It is no longer possible to dissociate knowledge from conscience, and progress from its consequences.

Let us look at some of the results of lack of foresight and of irresponsibility. Firstly, our atmosphere is slowly becoming polluted by an accumulation of carbon dioxide from industrial and domestic flues. The rate at which this gas has been absorbed by the atmosphere has increased constantly since 1900, and we are approaching the limit beyond which we must expect climatic changes due to partial filtering of the sun's rays. Granted, only a tiny fraction of the sun's rays will be absorbed, but even slight variations may well have serious repercussions on nature's extremely fine balance.

Another subject which should be thoroughly investigated is the increase of toxic and carcinogenic substances in our food and atmosphere. The metabolism of plants and animals depends directly on the composition of the air and the soil, and is bound to be affected adversely by such chemical by-products as petrol fumes, tars, dyes, and by boron compounds released by rockets and nuclear reactors.

In particular, the atmosphere is constantly being poisoned by atomic and thermonuclear explosions, and waste from nuclear power-stations. Slowly but surely, the composition of our natural environment is being changed by the introduction of complex molecules and radio-isotopes which were not previously present. What organic reactions to these processes are known, are known to be definitely harmful.

Again, when we consider man's effect on the other forms of life on earth we find that his actions have had the gravest results. Volumes would be needed to list all the disasters caused by human interference with nature. Whole species have been wiped out, deserts created, fertile areas eroded, or so altered as to disturb the whole balance of nature. Fertilizers, insecticides, and antibiotics, by which some of the worst damage can be repaired, are but substitutes, and in most cases they are used quite recklessly and with complete lack of consideration of any but the most immediate effects.

Many people have denounced the way in which we squander the earth's natural resources, and yet we continue to waste valuable substances at an alarming rate. At present trees are being cut down in such numbers that all available timber will be used up in fifty years' time. Even if we should discover ways and means of producing synthetic substitutes we cannot alter the fact that trees are essential for preserving the oxygen balance in the air. It is imperative that we maintain the essential balance between animals and plants, a balance that we have for far too long chosen to ignore.

There are many 'plans' to change different areas of the world. The implementation of these plans may strike us as marvellous, though we should, in fact, be most chary of them. Large-scale interference with nature is bound to have the most serious consequences, including changes of climate and hence of winds and rainfall, the creation of inland seas, and the melting of polar ice. Such alterations may bring local benefits but they may have adverse effects on entire continents. We need a new science which will make a specialized study of all the possible side-effects of such actions.

Finally, the first space-travellers must be launched with infinite caution. Once man lands on other planets, he will meet conditions quite other than those to which he is adapted. If micro-organisms from these planets are brought back to earth we might easily succumb to them, for we may have no natural resistance against them.

Scientific progress has reached the point where it can abolish all the scourges which have afflicted mankind since prehistoric times. But we must not use our new powers unthinkingly or immorally, for they are so vast that they can lay the whole earth waste.

There is urgent need for a new conception of science, based firmly on forethought and morality. Let us hope that before long the universities will establish chairs for investigating man's effects on nature. Such research would be infinitely more valuable than scientific work in specialized fields, carried out with complete disregard of the consequences.

This new approach would open the gateway to a 'meta-science', i.e. to a scientific philosophy, with a practical scope far greater than that of modern physics. For, notwithstanding all the marvellous discoveries on which it is based, modern science has every reason to be humble in face of its own shortcomings.

Conclusion

'Men are enclosed in their ignorance as
in a prison with slowly receding walls.
Unable to see beyond, they marvel at the
vastness of their mansion without ever
suspecting the existence of an infinite world
outside.'

ALFRED FRANKLIN

THE surrounding Universe, of which we are an integral
part, holds immense mystery. The very fact that during
the past few generations we have been able to explain so
many of its wonders is an indication of how very much
farther we have still to go. What we know already will be a
help, but we must also devise totally new methods of
discovery and exploration, methods which are certain to
arise from the mental evolution of our species.

The chief obstacles in our path are anthropocentricity,
uninquiring apathy, and every kind of prejudice. The
future belongs only to the inquiring and the enthusiastic.

There is a strange contradiction between our recent

succession of fundamental discoveries, and our continuing disbelief in the world as an inexhaustible source of revelation. Material and biological exploration of the cosmos have only just begun. Exploration of animal life is in its infancy; psychic exploration has hardly been touched upon; and the boundaries of the mathematical world have been glimpsed only by dreamers. And beyond all that we know, devise and sense here and now, is all the unknown, infinitely transcending the finite. So many other new discoveries lie ahead of us!

Why should we imagine that we know almost everything when we base the supposition on a frail edifice comprising what we have learnt in a few generations?

Has any conquest of knowledge been more than a grain of sand which, collected with all the others in the course of a few millenia, has enabled us to build a beautiful sand-castle on the shore of the ocean of unknown? Surely all the sand stretching away in both directions till it is lost beyond our horizons could be used to build a thousand million castles? Is humanity the only child on this sea-shore? Is there not a whole host of other children occupying themselves in the same way, each alone, scorning his neighbours, but at the same time spying? Who can say to just what limits the giant universal breath reaches? Billions of worlds gravitate. Matter is not the only constituent of the Universe. It must be realized that all around us lie possibilities beyond comprehension. Space and time are perhaps only dots in an infinite order, mightier than any we can conceive.

What transcends all these questions, however, and must lead us far towards the partial answer for which we hope, is the very limitation of our human life and knowledge. Here below there is nothing, but far beyond all is possible; an infinite possibility too rich and unexpected for us to imagine. Man, the world, and all our knowledge form a mere comma in a vast encyclopaedia which we have to decipher painstakingly. Will the future turn for us whole volumes, or merely a few pages?

Obviously, the continuing mental ascent of the human species can be predicted for a long time to come. The human intellect will seek still higher for the open country it has sought since birth. There is no guarantee, perhaps only a faint intuition, that we shall ever reach the goal. Perhaps, again, each living species in the universe has its own mountain—some huge, others only hillocks. Perhaps we are among those with only a limited height to scale, who will never see even at the peak more than a limited panorama. Or are we among those who confront the Himalayas, standing as yet at the foot of the mountains